Arbortext 102:
Best Practices for Creating Arbortext Styler Stylesheets

An Arbortext Monster Garage Book

Elizabeth Fraley
Single-Sourcing Solutions, Inc.

Arbortext 102: Best Practices for Creating Arbortext Styler Stylesheets

An Arbortext Monster Garage Book

Published by Single-Sourcing Solutions, Inc.

This book was written in DITA in Arbortext Editor and published using Arbortext Styler.

Single-Sourcing Solutions

PO Box 62122, Sunnyvale, CA 94088.

www.single-sourcing.com

twitter.single-sourcing.com

Electronic edition: October 2016

ISBN: 978-0-9976505-3-2

Books from Single-Sourcing Solutions

Arbortext Monster Garage Series

Arbortext 101: Best Practices for Configuring, Authoring, Styling, and Publishing with Arbortext

Arbortext 102: Best Practices for Creating Arbortext Styler Stylesheets

Arbortext 103: Best Practices for Configuring, Managing, and Publishing Content with the PTC Server (Coming Winter 2016)

TC Dojo Series

Adding Custom Actions to OxygenXML Framework

To anyone who has been told that, in this
century, publishing XML is hard (or expensive).

Contents

Chapter 1. Introduction

Topics Covered in this Chapter

- ♦ The Single-Sourcing Solutions approach
- ♦ Applicable software versions
- ♦ Acknowledgements
- ♦ About the author

This book is about Arbortext Styler, a graphical user interface (GUI) that facilitates the development of stylesheets for use with Arbortext publishing tools. It is remarkably successful at its task.

XML publishing has essentially three components: an XML document, a stylesheet, and a rendering engine. Stylesheets contain rules that define how to format content for display. That display—whether it is a physical book, a PDF, a website, or the text of an error message reference page—depends completely on the stylesheet and the rendering engine.

A rendering engine takes the XML document and the stylesheet and, following the rules laid out in the stylesheet, transforms the content contained in the XML document into the display format required.

Stylesheets have a *one-to-one relationship* with document types. They also typically have a one-to-one relationship with a display format. However, the same stylesheet can support any number of XML documents. Any XML document that is based on that same document type can use a stylesheet that is tied to that same document type.

The typical XML publishing process looks like this:

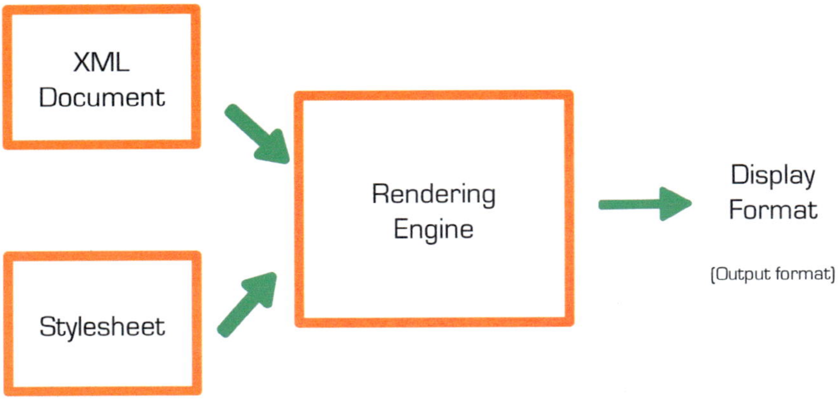

Every XML publishing system works this way. The rendering engine can be FrameMaker, the CHM compiler, Arbortext, or something else. Think of the "Stylesheet" as the .xsl (or XSL-FO) stylesheet, or the FrameMaker template, or the Arbortext Styler .style file.

You always have source documents. You always have something that provides the instructions about how to format something when it's tagged a certain way. And you always have something that turns one into the other. A binary PDF. A RoboHelp file. A Word document. An HTML file. Or simply another XML or text document.

Arbortext Styler stylesheets, however, have a *one-to-many relationship* with display formats. The Arbortext Styler publishing process looks like this:

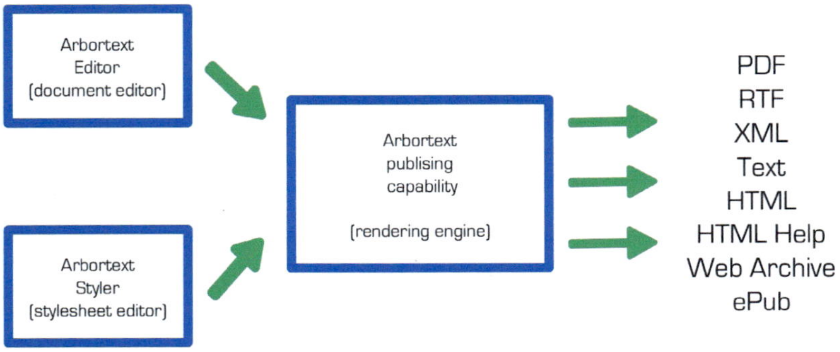

The same Arbortext Styler stylesheet can support multiple XML files *and* multiple outputs at the same time.

Since the company launched in 1991[1], Arbortext has been trying to address the problem of how to make new stylesheet developers productive quickly. They've had many variations over the years. Once they moved to incorporate the Arbortext Advanced Print Publisher as the primary rendering engine, Arbortext Styler became a powerhouse both in terms of capability and in skill acquisition for new users. Its unique graphical user interface means that no one must become a programmer simply to format technical content.

The Single-Sourcing Solutions approach

This book walks you through our approach to stylesheet development for new users of Arbortext Styler tested over sixteen years of using the product. We have worked with clients who had us develop stylesheets for them and with clients who wanted to build their stylesheets themselves, with guidance from us.

Our approach produces a stylesheet that is 85% compatible with layout for requirements typical of most technical publications departments. The lessons learned by following our process highlights the building blocks needed to complete that last 15% that is unique to your company's publishing needs.

This is the method we use when we develop stylesheets for our customers. It's also how we teach them to build stylesheets. It's a straightforward method that makes things easy to identify, locate, and track down later.

Our process begins with a stylesheet to support print output because it is the most familiar output from the most common paradigm (desktop publishing) and illustrates some important concepts in stylesheet design with Arbortext Styler.

This book will provide you with a blueprint for creating stylesheets that give you the best chance for long-lasting, low maintenance publishing in Arbortext. By following our process, you will know how to develop a fully functional stylesheet that produces content customized to your unique needs—and you can get it all done in a matter of hours.

---

1. See the website via the wayback machine at wayback.arbortextuser.org

Applicable software versions

This book is current as of Arbortext 7.0 M050 (released 8 December 2016) and assumes PTC Server 11.0 F000 or higher (released 17 December 2015). We are mindful of crossover issues with Arbortext Advanced Print Publisher 11.1 M030 (released 22 June 2016).

> **Expert tip!**
>
> With the M050 release PTC repositioned Arbortext under the Servigistics product line. It is now officially known as "Servigistics Arbortext."

This book was written in DITA using Arbortext Editor and published for print, PDF, ePub, and MOBI using Arbortext Styler.

Our examples assume a local installation of Arbortext Styler. Where we discuss print stylesheets, we assume those stylesheets use the Arbortext Advanced Print Publisher (APP) rendering engine for print publishing.

About the code examples

DITA is the doctype used in the examples, but the principles remain the same regardless of the doctype you use.

Throughout this book, the examples include the DITA files from this book and from Arbortext 101 for illustration.

For the exercises, I use copies of the samples provided by Arbortext out of the box with a few additional DITA topics. You can make your own example source files or you can download additional DITA sample content from my website: http://102samples.single-sourcing.com

Acknowledgements

Arbortext knowledge has always been tribal, but at Single-Sourcing Solutions, we have always done everything we can to share our knowledge with the world around us.

I want to thank Janice Summers, my secret weapon. Single-Sourcing Solutions has succeeded greatly due to her caring friendship and coaching. She has a particular knack for making the complex simple. She tested everything in this book and has fought for every inch of her Arbortext knowledge. Without her, I could not have done half as much as I have and I'm beyond grateful to her.

Thanks to Lori Meyer, my editor and friend. She's always encouraging and is an outstanding editor. I have always known I needed a good editor and she makes me better.

Thanks to all of our customers for encouraging me to write this book and make it available for anyone who wants to learn how to use Arbortext to their best benefit.

Thanks also to the many competitors we have that inspired me to write this book in particular. Customers come to us over and over with poorly written stylesheets that depend on old technology. We feel their pain. At Single-Sourcing Solutions, we believe you are 100% capable of doing anything that we do and we want you to know how. I hope that this book gives you the capability to judge the work others have given you.

About the author

Elizabeth Fraley is a serial entrepreneur. She's founded two companies, sits on the boards of three non-profits, and is constantly coming up with new ways to share knowledge in the technical communications and content industries.

She works as a Single-Source/XML Architect/Programmer and Mentor. She has worked in industries ranging from high tech to government, at companies of all different sizes (from startups to huge enterprises). She advocates approaches that improve organizational efficiency, productivity, and interoperability. She takes an apprentice to journeyman approach with her customers, holding their hands and enabling them to stand on their own two feet. She knows that her customers want to do it themselves; they just don't want to do it alone.

She constantly looks for ways to make tribal knowledge...not so tribal. She presents regularly at international, national, and regional conferences related to content, information architecture, and technical communication. Her appearances and publications are consistently well received by her peers. She is in the process of launching a new program designed to assist technical communicators grow their professional business and political skills because those skills are as important as technical prowess for success.

She has created two community-driven webinar series (the Arbortext Monster Garage and the TC Dojo), three YouTube channels, and an Arbortext On Demand Vimeo channel. She has an Arbortext Community Voices podcast, and runs the Arbortext Meetup Group, LinkedIn Group, and Facebook group. She writes blog posts and publishes papers that enable the people in her communities. She developed the first mastermind series for technical communicators and, as president of Single-Sourcing Solutions, saved an internationally known Arbortext code archive from deletion. She's the founder of TC Camp, is the only unconference focused on the technical writing community and the people who support them.

Chapter 2. The Arbortext product suite

Topics Covered in this Chapter

- ♦ Understanding Arbortext Publishing
- ♦ Purchasing Arbortext Styler
- ♦ Licensing Arbortext Styler
- ♦ Installing Arbortext Styler

The Arbortext product suite offers solutions at every stage of information development, from defining, authoring, and illustrating to managing and delivering content. Arbortext is an end-to-end product information delivery system that automates the authoring and publishing processes for the delivery of high-quality, tailored product information in the form of operator manuals, service documentation, and eLearning courseware.

Deployed as a complete integrated product information delivery system, the components of the Arbortext Product Suite are built and tested together, to ensure out-of-the-box deployment with low risk, easy upgrades, and a low total cost of ownership.

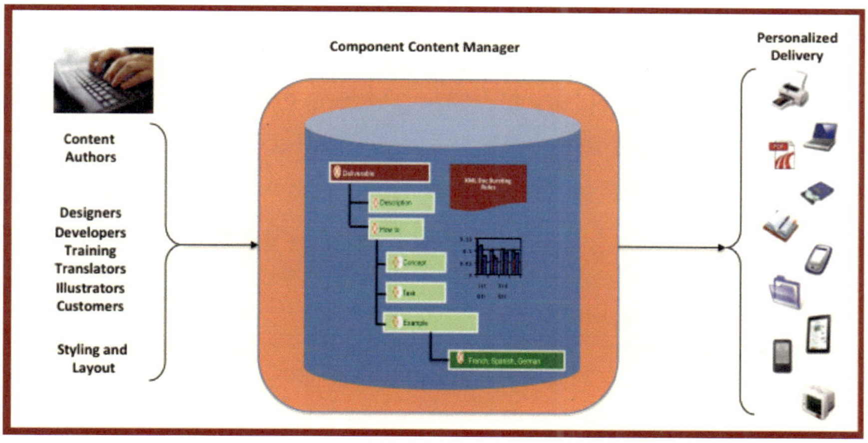

Most customers purchase a subset of the Arbortext Product Suite. Everyone needs to create content (authoring), design the look and feel of finished output (styling), create final publications (publishing), and manage all of those content units (content management).

A typical Arbortext deployment includes the following products in the Arbortext suite:

Arbortext Editor (authoring)	A familiar word processing interface that makes it easy for authors to create and edit structured XML and SGML content for reuse and automated publishing.
Arbortext Styler (styling)	Enables publication designers to create multi-channel XML-based stylesheets for automated publishing in a simplified user interface.
Arbortext Publishing Engine (publishing)	Automatically extracts XML and SGML content from file systems or content management systems, assembles that content for different audiences, and publishes the assembled content automatically to print, web, PDF, Microsoft Word, HTML Help, and other media formats.
Arbortext Content Manager (content management)	One brand name for the PTC Server. It is an integral component of the PTC Dynamic Publishing System that provides a single source of information, maintains control at every component level, preserves component relationships, and provides configuration management capabilities to enable organizations to optimize their publishing process.
Arbortext eLearning Library (training)	This library of training videos contains material dating back to Arbortext 5.2. The video content is exactly the same as the official PTC in-person training course material, except that it is broken into one-hour segments that make it easy to view and work into your daily work schedule.

Understanding Arbortext Publishing

Three concepts are essential to understand Arbortext Publishing:

1. Arbortext Advanced Print Publisher (APP)
2. Arbortext Styler
3. Arbortext Publishing Engine (APE)

Before you can truly understand Arbortext Styler, you have to understand the rendering engine (APP) and the publishing engine (APE). They are discussed in the following sections.

Arbortext Advanced Print Publisher

Arbortext Advanced Print Publisher (APP) is a desktop publishing application that originated as a high-end typesetting tool. APP is also the primary rendering engine used by both Arbortext Styler and Arbortext Publishing Engine.

APP Desktop is a powerful print publishing tool used in environments where there are complex layout requirements and rapid production times. It has a WYSIWYG component and an interactive touch-up tool that allows publishers to build pages manually and immediately visualize the result of formatting and style changes. Although it sounds like the desktop publishing (DTP) applications we are familiar with, it is far more complex. Due to the huge emphasis on automation features and the SGML-like syntax used to describe layout and formatting, APP is considered an advanced formatting application.

With the Arbortext 5.4 release, APP became the primary print rendering engine for both Arbortext Styler and Arbortext Publishing engine. Arbortext Styler stylesheets can take advantage of the power of APP through several mechanisms, including Perl and JavaScript programming.

> **Expert tip!**
>
> Arbortext users distinguish the desktop application from the rendering engine by using *APP Desktop* to refer to the desktop application, and *APP* to refer to the rendering engine contained in both Arbortext Styler and APE.

Arbortext Publishing Engine

Arbortext Publishing Engine (APE) is a server publishing tool. When configured for use with Arbortext Editor, users can publish without requiring a license of Arbortext Styler, APP desktop, or any other rendering engine to publish any supported output format.

Like Arbortext Styler, APE uses APP as its rendering engine for print publishing but it can publish more than print output. It also includes the capability to publish all the other formats also supported by Arbortext Styler:

- Arbortext Editor
- PDF
- Print Preview
- HTML
- ePub
- HTML File (includes XHTML and HTML5)
- HTML Help
- Web archive
- RTF

Technically, as of this writing, APE users still have a choice of rendering engines for print output:

- Arbortext Print Publisher (APP)
- Formatting Output Specification Instance (FOSI)
- XSL-FO[2]

With the release of 7.0 in August 2015, PTC announced that support for the FOSI and XSL-FO engines have been moved to sustained support. They are no longer developing new features or fixing bugs.[3] We recommended always using APP for print output.

Expert tip!

Even though you can choose a different rendering engine, we recommend always choosing APP for your print output. Do not accept a stylesheet for use in your Arbortext environment from a consultant that is not APP-based. You don't want to start with a stylesheet that is based on technology that has reached its end-of-life.

---

2. Arbortext Styler has an XSL-FO rendering engine but does not create .xsl stylesheets. If you specify the XSL-FO engine, it will use the FO engine to drive formatting and page layout from the .style stylesheet you create in Arbortext Styler.
3. See the announcement: http://fosi-eol.single-sourcing.com

Arbortext Styler

Arbortext Styler is a stylesheet editing application. When Arbortext users talk about Arbortext Styler, they always refer to the *stylesheet editor*.

In reality, an **Arbortext Styler** license unlocks access to three tools:

- Arbortext Editor (a document editor)
- Arbortext Styler (a stylesheet editor)
- A rendering engine (the capability to publish output)

A single license of Arbortext Styler authorizes one desktop user access to everything inside the red box:

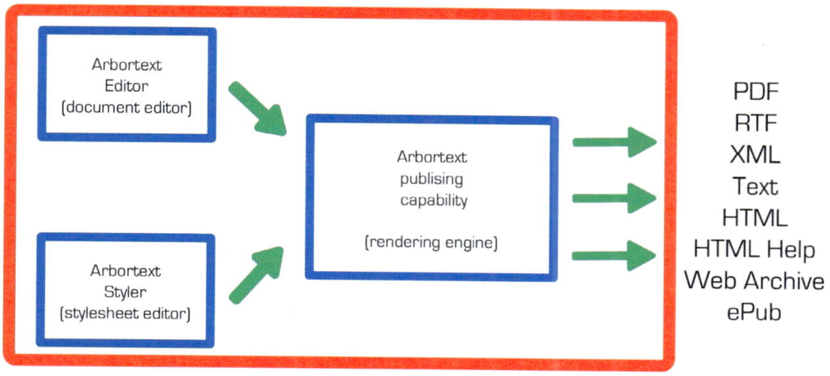

Arbortext Styler provides publishing capability because stylesheet designers need to be able to test their stylesheets. In order to create, maintain, and test stylesheets, all three components are required. You must have a document that you can use as the sample input for the process; you must be able to edit the stylesheet; and you must be able to render output to verify that the stylesheet does what you expect.

Because Arbortext Styler is a GUI application and not a programming interface or text editor, Arbortext Styler does not require stylesheet designers to be programmers. Any Arbortext Styler stylesheet can be used as a module for another stylesheet, adding flexibility while reducing maintenance overhead and development time.

All of the rendering engines available to APE are available to Arbortext Styler. All of the output formats supported by APE are supported in Arbortext Styler.

Arbortext Styler provides the interface for stylesheet designers to create Arbortext stylesheets that support publishing to multiple outputs.

Once the stylesheet is ready, Arbortext Styler stylesheets can be distributed to APE for publishing by anyone who has an Arbortext Editor on their desktop.

In this book, Arbortext Styler refers to the GUI application that helps designers define rules for formatting the elements in a doctype and provides push-button publishing capability.

Normally, I would not explain an application's user interface. Most of us are confident at exploring a UI on our own. However, a couple of items here bear pointing out because I'm going to have you hopping between them. The stylesheet editor looks like this:

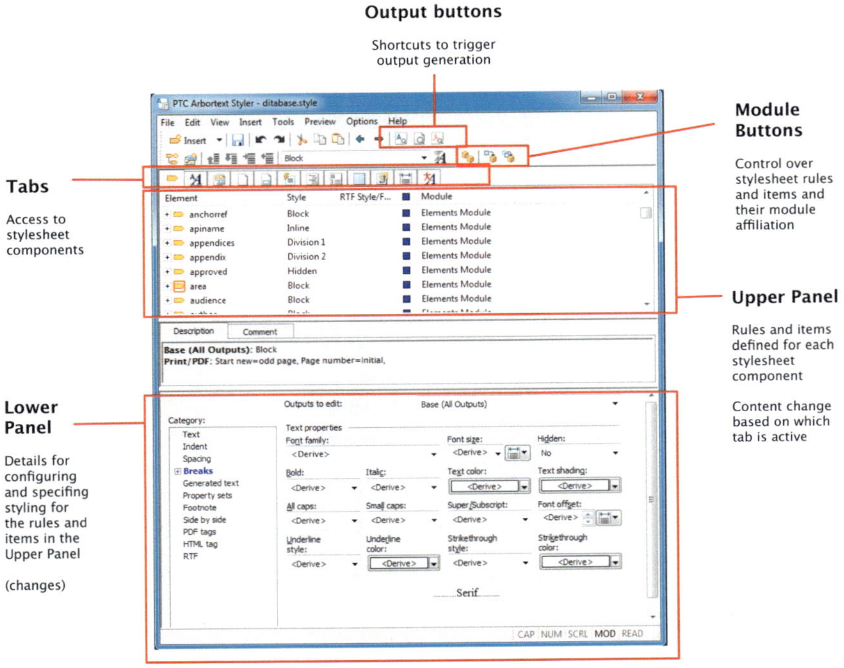

The **output buttons** provide a shortcut to generate output for the document editor, a print preview, or PDF. When you're working with print stylesheets, these buttons make it easy to see the result of your changes right away.

The **modules buttons** give you control over where each rule or configured style item (like a page set or a cross reference definition) will be stored in the module hierarchy. You can also access these features through the File menu, but the buttons make it easy to move rules around.

These three buttons provide quick access to functions you will use *all the time*:

- Open Modules Window 
- Move to module 
- Copy to module 

The **tabs** give you quick access to each of the major stylesheet components. The tab arrangement gives organization to your approach to stylesheet work. As you move left to right, the components become more intricate. The tabs are defined below. Tab usage will become more clear as we walk through examples later in the book. For now, it's important to understand that each tab controls a different stylesheet component and changes what you see in the upper panel.

- **Element Rules** – the list of elements and contexts that can appear in any source document
- **Property Sets** – reusable style definitions
- **Page Sets** – a list of page sets
- **Page Types** – the number of page types for each page set. Every page set contains a number of page types
- **Page Regions** – regions associated with each page type. Every page type has regions on it like headers, footers, and a body
- **Generated Content** – reusable definitions for constructed content
- **Tables of Contents** – reusable TOC units and their definitions
- **Indexes** – reusable Index units and their definitions
- **Custom Tables** – definitions that tell the stylesheet how custom tables should be constructed from non-traditional table markup (for example, CALS)
- **Cross References** – a list of cross references and their defined behavior
- **Size** – a list of reusable variables that specify dimensions and sizes

- **Combined Fonts** – a way to handle complex Unicode character and font display requirements within the same document (an advanced topic)

The contents of the **upper panel** change depending on which tab is active. The upper panel includes the list of stylesheet items that can have styling applied to them or which define style properties for use in other tabs. Think of the **upper panel** as having the rules.

The **lower panel** displays the property definitions of the rules selected in the **upper panel**. For element rules and property sets, content of the lower panel also depends on which category is selected in the **category pane**.

The Arbortext Styler GUI application removes stylesheet programming from the hands of a few and puts it into the hands of every Arbortext user. As you can see in the **center panel**, Arbortext Styler stylesheets can support all of the supported output formats *in the same stylesheet*.

> **Expert tip!**
>
> Even though Styler has the capability to convert a FOSI to an Arbortext Styler stylesheet, we recommend starting fresh whenever possible. FOSI and APP approach the world so differently that it is often simpler to start fresh. In addition, it's a perfect time to re-address your style and formatting choices. Stylesheet migration will give you the opening you need to refresh your layout to meet modern user experience best practices.

A statement about standards and best practices

Whenever you talk about XML publishing, you are talking about standards. XML is a standard. DITA, DocBook, and S1000D are all standards that define how to construct valid, well-formed instances of documents that conform to their specific vocabularies.

The files you create are source documents: The archive format you save, store, version, and lifecycle. The vocabularies might have changed over time, but XML documents are ASCII documents; and, unlike the many Word for Windows documents I still have on old 3.5-inch floppy disks, my XML documents will be as readable tomorrow as they are today. (At least as long as I have a way to access them. Bye bye floppies.)

Even though I can open an XML file that was created 20 years ago, it's highly likely I can't publish it in a nice display format today without some significant work. No one knew what it would take to publish to iPad 20 years ago because it didn't exist. No one knows what new display requirement will come 20 years from now. If I want to publish a 20-year old XML document to an iPad today, I will need to create a stylesheet that I can use to publish with today's technology.

Publishing technology is always changing. It changes to fit new applications and new display devices. It advances as fast as display options appear in our hands. It's very difficult to limit the options for future publishing requirements when we haven't even begun to guess at where they might go.

Will online help become truly integrated where the troubleshooting steps become live actions (not just instructions for what to do and verify)? Will a reader be able to check the tolerance of a nuclear reactor? No one knows, because right now we don't have software or display technology that allows this behavior. But by using standard formats for content, as well as best practices for publishing, you will get the most bang for buck for your hard work for years to come. Not everything will be reusable, but by separating content from style you will have a nimble process in place to handle the challenges of the future.

In addition, because we are all trying to produce the best product for our customers today, we know how we must publish and display our information to provide an optimal user experience. XML is critical if you want to be able to rapidly meet changing technology demands.

However, I don't believe that I should limit my publishing capability because the path uses a bicycle instead of my feet. My customer's needs come first—and I'll use whatever technology, tool, or option produces the best result for them. I know that as long as I can access the content, technology to transform will be available to me in the future to transform it. And I know I'll have to recreate the publishing path when the requirement arrives.

There are standards for some publishing technologies, but If you need a feature that a standard doesn't support, you'll likely find the support in an extension provided by your tool vendor. I say use it. It is most critical to adhere to standards for content. For style, you need to address the needs of your customers today. By adhering the best you can to the standards for style, it is more likely you will be able to reuse most of the hard work in creating the stylesheet, but style is the wild card, so don't count on it. If you are going to diverge from a standard, do so with the style. Do not gamble with the content.

Expert tip!

Don't worry that you're not conforming to standards in the technology you use to transform your archive source to your presentation source. Instead, be more concerned that the starting point and ending points provide you the best chance for survival of your content and the support of your customer's needs. The path is guaranteed to change.

Purchasing Arbortext Styler

Arbortext Styler is currently available for Windows only. For details, check the Software Compatibility Matrix for the version you're looking to deploy. This matrix is updated with every release and is available with the rest of the documentation set on ptc.com.

PTC does not sell directly to users over the Internet, but they do have a large channel partner network of value-added resellers. Partners have different specialties and are assigned to geographic regions. You can find a reseller partner on the PTC website.

Expert tip!

We highly recommend that you select a qualified Arbortext partner for your purchase as well as for your service partner. These are the partners whose primary focus is the Arbortext product suite. They live it every day and know the tools inside out. However, even if PTC assigns you to a different reseller partner for purchasing purposes, you can still use one of these elite partners for services and consultation. You can get good advice from a services partner who has extensive experience with the Arbortext product suite.

To find a list of Arbortext qualified partners, look at the "Everything Arbortext in One Place" article on Hub Pages: http://everything.arbortextuser.org

Licensing Arbortext Styler

Anyone who will be authoring, editing, or working with Arbortext Styler stylesheets requires a licensed copy of Arbortext Styler installed on their Windows computer.

Arbortext Editor comes with the free version of Styler. This version will only support stylesheets for use inside Arbortext Editor. It does not support other outputs (PDF, HTML, etc.) nor can it be used for publishing.

The full version of Styler allows you to develop stylesheets for all supported formats. The full version is licensed concurrently. Concurrent licenses float and can be shared among everyone who will be working on the stylesheets. Only one person can use the Styler license at a time.

In this book, Arbortext Styler means the full version.

Arbortext stylesheets can be deployed to the Arbortext Publishing Engine (APE) for server-based publishing. Authors using Arbortext Editor can configure their environment to publish with APE so they can use the stylesheets without requiring a license of Styler.

Arbortext Content Manager (ACM) is the component content management system and source repository. ACM can be configured so that content can be published with APE directly from inside the ACM web client without requiring a license of either Arbortext Editor or Arbortext Styler.

Getting trial licenses

Trial licenses of **Arbortext Editor with Styler** are available from the ptc.com website. The trial license is for both products (Editor and Styler). Licenses are valid for 14 days *from the date of download, not the date of installation*. It can take up to two *business* days to get your trial license from PTC, so be sure to take that into account before you download your trial on a Friday afternoon. Only one trial license is available per MAC address, so you need to remember that too.

If you require a longer trial, ask a channel partner. They may be able to help. Be aware that channel partners are required to have a valid business case to get an extended trial license issued by PTC. Channel sales teams must provide expected sales close dates and overall deal numbers for PTC to authorize an extended trial license.

Installing Arbortext Styler

Arbortext Styler is installed like any other Windows program. It is part of the Arbortext Editor installation in the **Program Files** directory of your Windows computer. If you already have Arbortext Editor installed, all you need to do is install the Arbortext Styler license to activate Styler on your computer.

Installation takes about 30 minutes, including licensing. Once the applications are installed, you can use Arbortext Styler right out of the box.

I assume that you've followed best practices for setting up your Arbortext applications and configuring your environment (the topic of *Arbortext 101*). While I will layout the structure for you to use for the step-by-step procedures described in this book, I won't go into why I chose these structures or provide alternative configurations. I will assume you have already configured your environment using best practices.

This book outlines the best practices for setting up your Arbortext stylesheet for best results using it today and maintaining it tomorrow.

Chapter 3. Your first stylesheet

Topics Covered in this Chapter

- ♦ Prepare your environment
- ♦ Open a sample document
- ♦ Create the stylesheet

In this chapter, I'll take you through the basics of working with Arbortext Styler stylesheets. For best results, follow best practices for configuring your environment. That way, you will have flexibility for your first stylesheet and for every stylesheet that comes afterward.

The concepts presented here serve as the foundation for every stylesheet you will build. By the end of this chapter you'll understand how stylesheets come together, and you'll have a pattern for building more complex stylesheets and customizing existing stylesheets to meet your publishing requirements.

And you'll have created your first one.

I'm going to use the DITA as the doctype for my examples. The principles remain the same regardless of the doctype you are using.

Prepare your environment

Before you start working with Arbortext Styler stylesheets, it's important to set up your custom directory. If you've configured your custom directory structure to match the best practices I described in *Arbortext 101*, you need to make only a few changes.

Before you start working with Arbortext Styler stylesheets, it's important to set up your custom directory. If you've configured your custom directory structure to match the best practices I described in *Arbortext 101*, you only need to make a few changes.

If you haven't read *Arbortext 101*, I'm not going to delve deeply into the rationale behind why you should configure the custom directory this way. As long as you construct your custom directory the way I describe in this chapter, you'll be able to work through the rest of the concepts described in this book easily.

Arbortext Styler stylesheets depend on the custom directory. Configuring your custom directory well will save you time and will guarantee consistency in every stylesheet you create.

Arbortext Styler stylesheets use the following directories:

- graphics
- framesets
- doctypes
- stylermodules

Stylesheets can automatically insert graphics, like your corporate logo, that are commonly used or controlled. The same image appears everywhere. By having the stylesheet insert the logo graphic automatically, authors don't need to worry about finding the right one. When it comes time to release a new version of the logo, or to change the branding, if we replace the logo file in the **graphics** directory; the stylesheet will use the new one automatically the next time you publish. Any graphic defined in the stylesheet should be placed in the **graphics** directory.

Stylesheets can produce web output. By default, web output has a left-hand navigation with search, a top frame that has branding and provides access to other functions like bookmarking and index terms, and a center pane that displays the content of the main HTML content pages. The template used to construct web output is stored in the **frameset** directory.

Stylesheets are tied to doctypes. The **doctypes** directory is where Arbortext publishing looks first for stylesheets that can be applied to a specific doctype.

Stylesheets can use (and be used by) other stylesheets. The **stylermodules** directory is the place where Arbortext Styler will look first for modules referenced from a stylesheet.

Each of these directories must be located in your custom directory and must have these exact names. If not, Arbortext Styler won't be able to find the resources essential to the publishing operation.

Arbortext Styler comes with fully functional stylesheets for both DocBook and DITA. You can use these as a basis for your own stylesheets. By copying the out-of-the-box (OOTB) files to your custom directory, you get a head start. You can use the OOTB files when you create your first stylesheet. At the same time, by having copies in your custom directory, you are protected against unanticipated changes made when Arbortext upgrades the software and provides new versions of the OOTB files.

This protection can also be seen as an added benefit. When Arbortext first added support for DITA 1.2, they provided all the new topic types as well as support for displaying those types in their OOTB stylesheet. By replacing the old OOTB files with the new ones, you got formatting coverage for all the new elements for free.

In the rest of this chapter, you'll create a custom directory that matches best practices for Arbortext usage and stylesheet design. You will create copies of useful files to your custom directory where any new stylesheet you make can take advantage of them. When you're done, you'll have a solid foundation that provides the structure for you to experiment and tailor Arbortext Styler stylesheets to fit your requirements.

Let's Do It!

If you've configured your custom directory structure to match the best practices described in *Arbortext 101*, you can skip ahead to Copy .style files, page 26. If you have not already configured your custom directory according to best practices, make sure any fonts you need are installed on your machine and follow all the steps in this chapter.

Create the directory structure

Create all of the following directories:

1. Create C:\Arbortext
2. Create C:\Arbortext\custom
3. Create C:\Arbortext\custom\doctypes
4. Create C:\Arbortext\custom\doctypes\ditabase

5. Create C:\Arbortext\custom\graphics
6. Create C:\Arbortext\custom\stylermodules
7. Create C:\Arbortext\custom\framesets

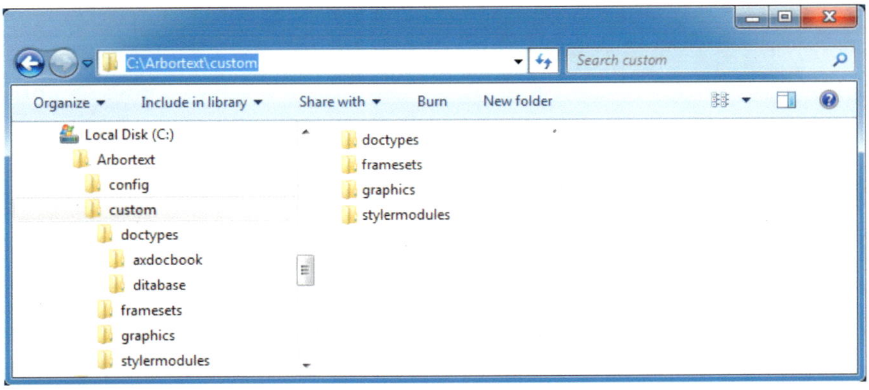

Identify your custom directory to Arbortext

Tell Arbortext the location of your custom directory by creating a Windows system environment variable named APTCUSTOM.

- **Variable Name**: APTCUSTOM
- **Variable Value**: %H%;C:\Arbortext\custom

Open Windows Explorer and right-click on **Computer**.

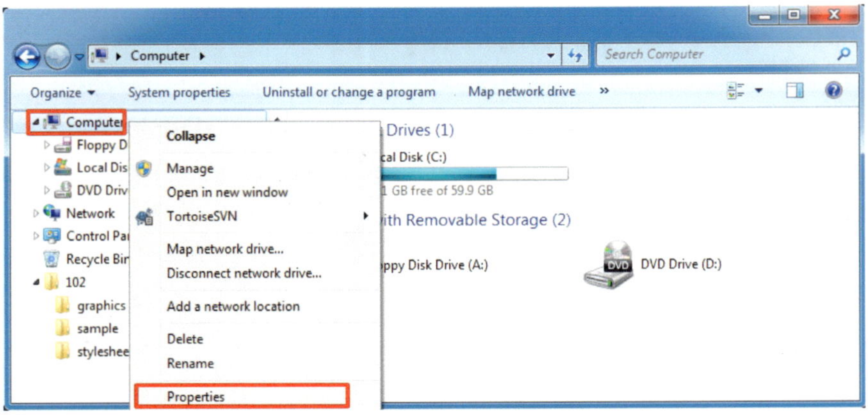

Select **Properties** to open **System** settings in the Windows Control Panel.

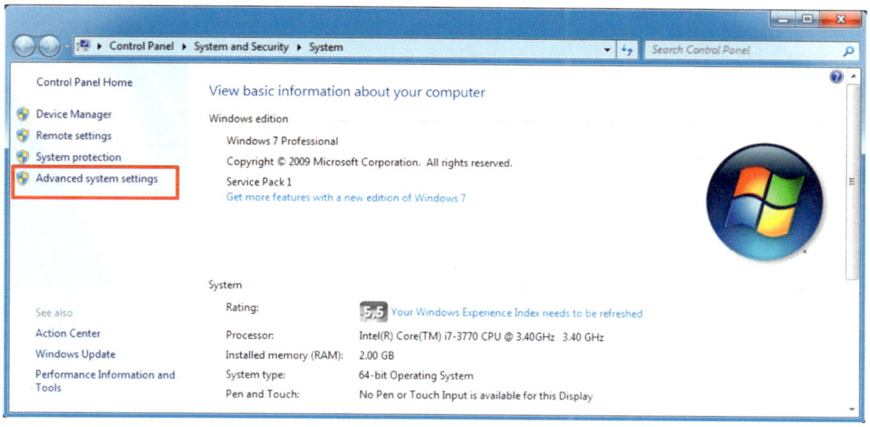

Click **Advanced system settings** in the left-hand navigation to open the **System Properties** window.

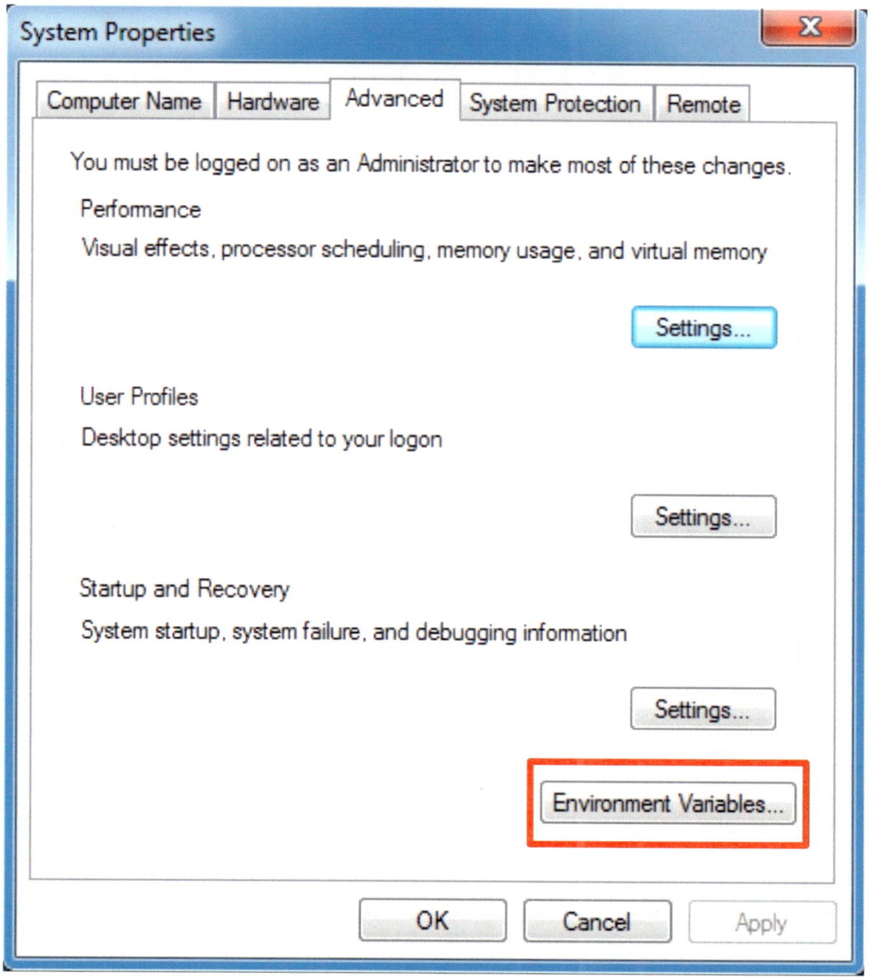

Click **Environment variables** to open the **Environment Variables** window where you can create the system variables.

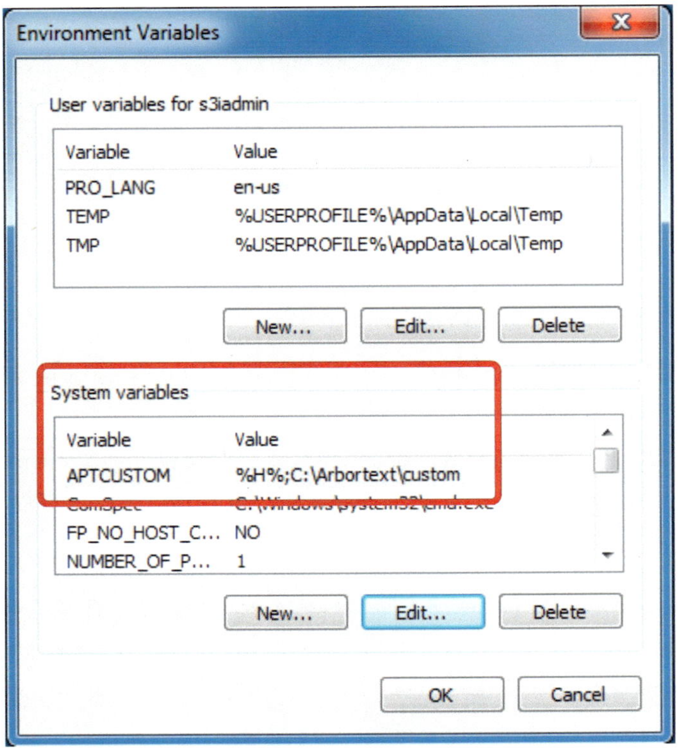

Add graphic files

Copy your company logo into the **graphics** directory.

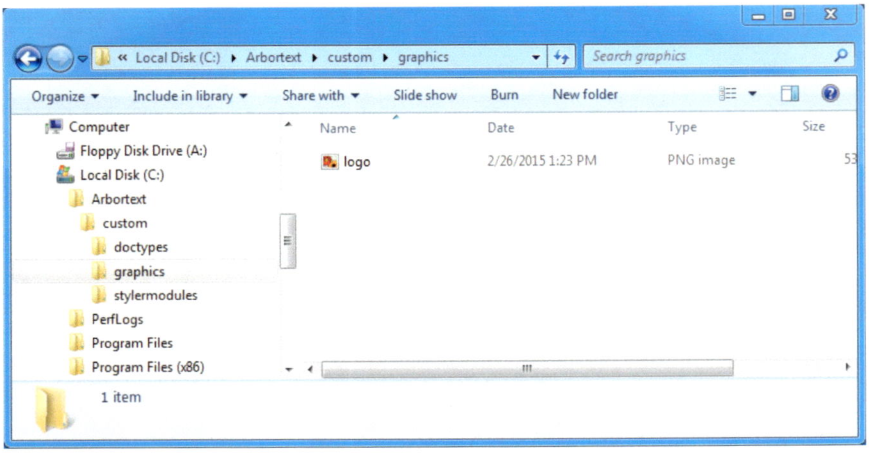

If you have any other graphics files that are tightly controlled and common across your publications (such as graphics for notes, cautions, and warnings) you can copy them here too. Be aware that this folder is not for *all* graphics. Graphics that belong to any particular document should be stored with the content they belong to. The only graphics that go here should be the ones that rarely change, that are the same on every publication, and that should not be changed without careful consideration and approval.

Add the default frameset template

Copy the framesets **default** directory from the install tree into the matching location in your custom directory.

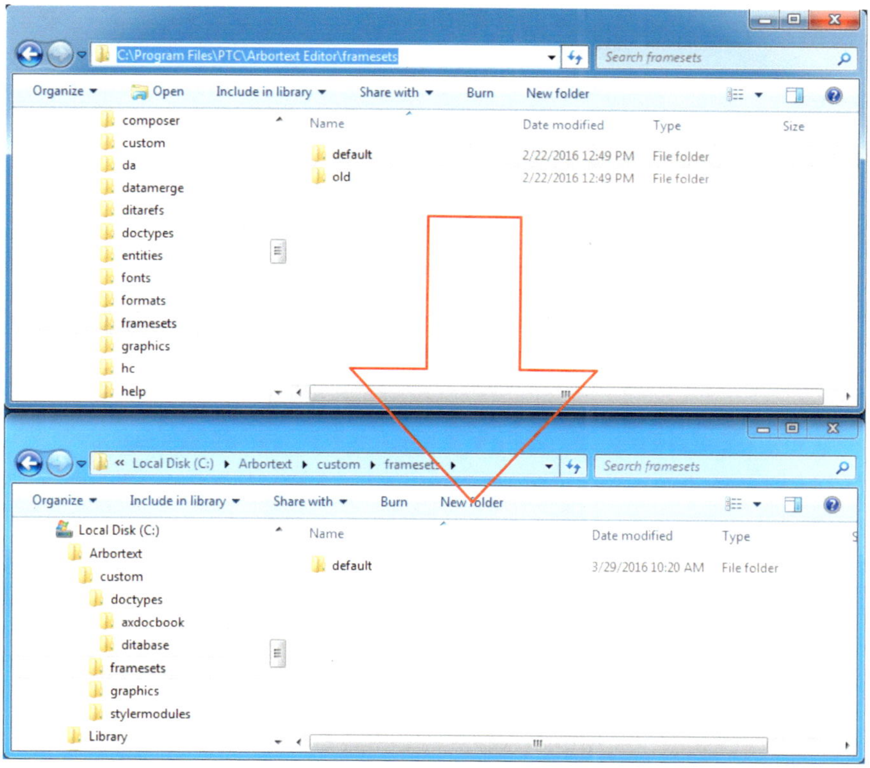

Copy .dcf and .pcf files

These two types of files affect how a doctype behaves in your environment. Copy both types of files for both doctypes into the matching folders in your custom directory.

Copy the default **ditabase.dcf** and **ditabase.pcf** files from the install tree to the **ditabase** doctype directory in your custom directory.

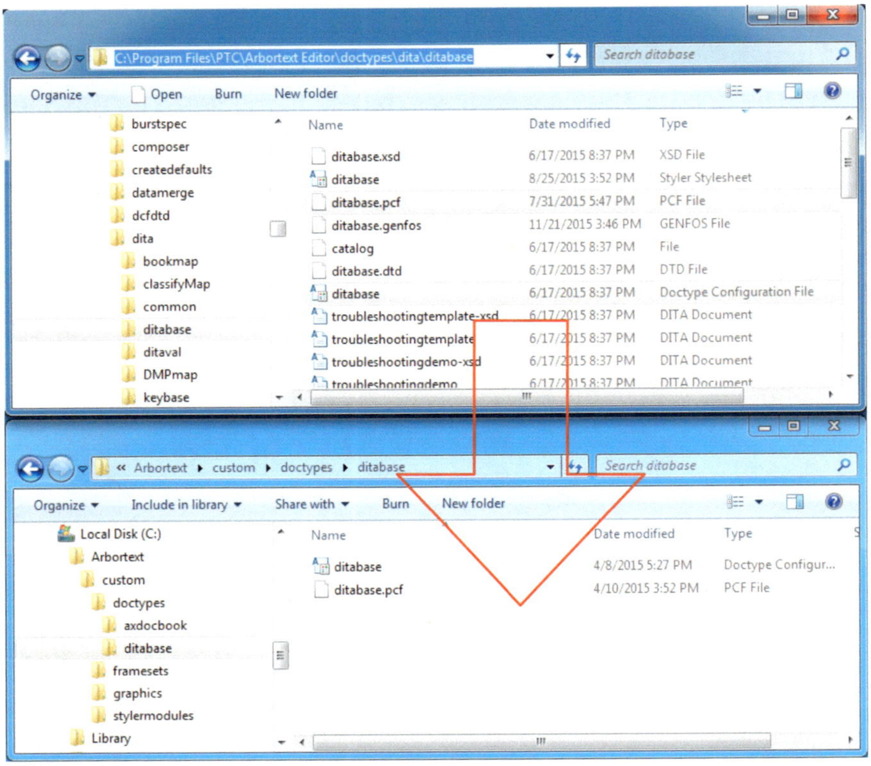

Copy .style files

For DITA documents, we want to use the DITA modules rather than the **ditabase.style** file found in the ditabase doctype directory.

By copying the **.style** files to the **stylermodules** directory, we are making the out-of-the-box, standard DITA stylesheet available as a module for future DITA stylesheets.

Navigate to the **stylermodules** directory in the install tree.

Copy all the **.style** files found here into the **stylermodules** subdirectory of your custom directory.

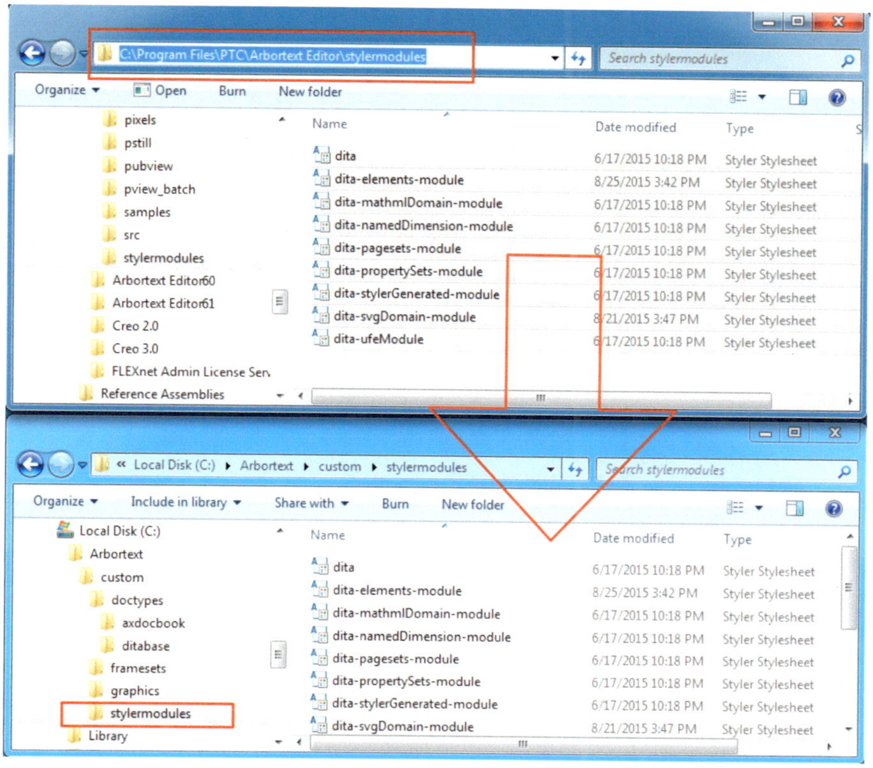

Now we have the out-of-the-box, standard DITA stylesheet and available to use as a module in future DITA stylesheets.

Expert tip!

If you set up your custom directory according to Arbortext 101, you need to do two things:

1. Delete the **ditabase.style** file out of the **ditabase** doctype directory.
2. Follow the procedures to copy the **.style** files in the **stylermodules** directory in the install tree into the **stylermodules** directory in your custom directory.

Fix permissions on copied files

Change permissions on copied files so that they are not marked **Read Only**. If you change permissions on the entire custom directory, you'll change all the files you copied all at the same time.

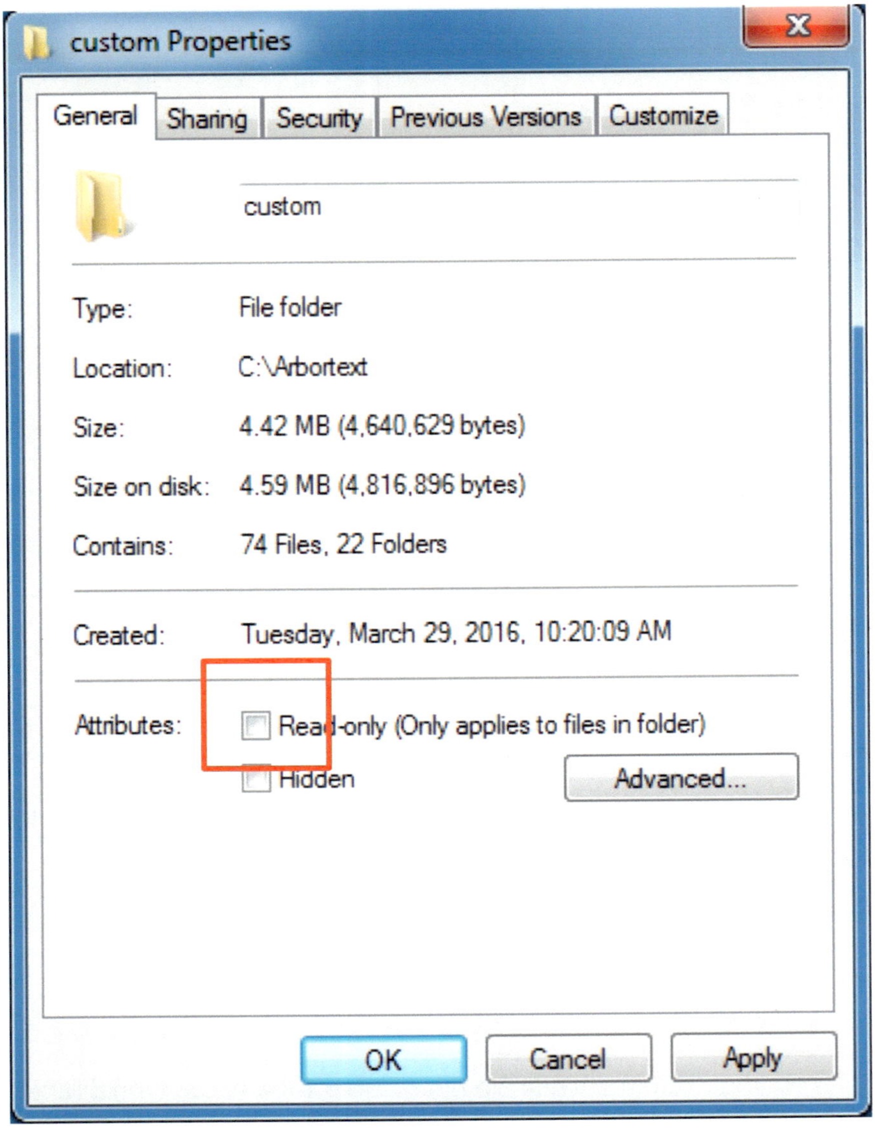

Make sure the changes apply to **this folder, subfolders and files**.

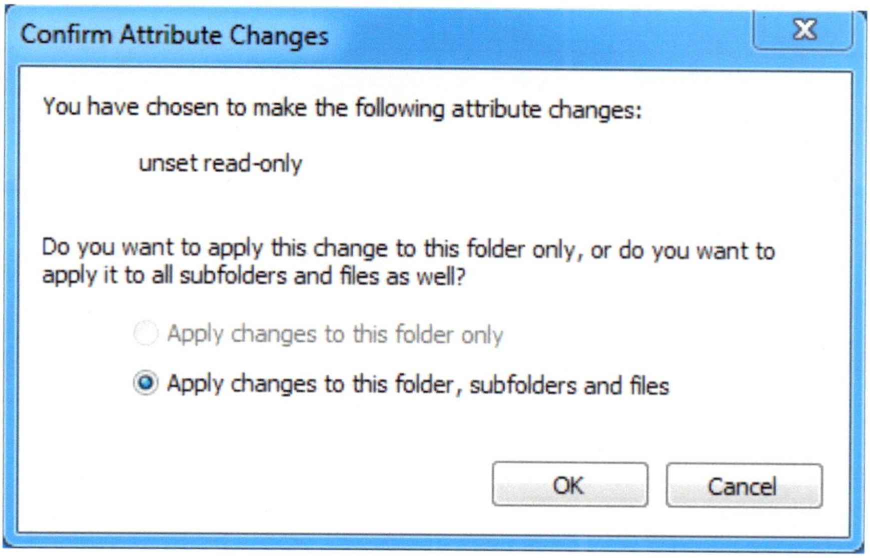

Open a sample document

Before you can open the stylesheet editor, you must have a document open in the document editor. The type of document open in the editor determines the kind of stylesheet you can work on. Arbortext provides a way for you to get a sample document easily.

When you go to **File→New...**, you get the **New Document** window which has three parts. The **Category** on the left hand side is the list of available document types. The **Type** pane on the right hand side is the list of root elements for new documents of that type.

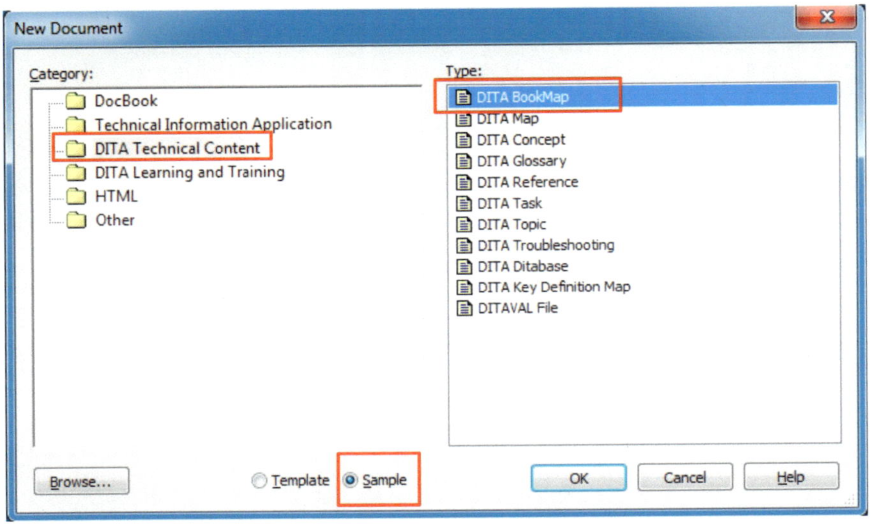

At the bottom of the **New Document** window are two options. If you choose **Sample**, Arbortext will create a sample file with content containing both text and images marked up with an assortment of tags found in a typical document of that type.

The DocBook category is a version of DocBook 4.0 that has been configured to work with Arbortext. The DITA Technical Content and the DITA Learning and Training categories are the OASIS standard DITA doctypes. The Technical Information Application is a DITA specialization created and maintained by PTC in response to typical customer requests. You can install your own doctypes and create a category folder for them. See the Arbortext Help Center for more information about installing custom doctypes.

Once you have a sample document, you can open the style editor to work on stylesheets that match the document category. A DITA sample document will let you work on DITA stylesheets and a DocBook sample document will let you work on DocBook stylesheets, for example. You can create a new document sample any time you like by choosing the **Sample** option and clicking **OK**.

The document that is created when you employ the **Sample** button is a sample instance of the document type you chose. You can change it to create representative examples of the content you create. You can create as many samples as you want. Changes to the document in the editor do not change the way samples are created.

If you prefer to use your own sample, you can open your sample instead of creating a new sample document on the fly. The capability for sample documents provided with the document editor is there for your convenience and the convenience of new users learning how to use the different document types.

Expert tip!

When you create a new sample **DITA Bookmap**, you are only creating only one document, the bookmap. You are not creating all the topics, tasks, concepts, and references that are referenced in the bookmap. If you want to create a complete duplicate set of the bookmap and all of the topics, you create copies of the files provided by Arbortext. All of the sample files you get through the **New Document** dialog box are in the **ditarefs** subfolder in the install tree.

Let's do it!

Copy the **ditarefs** folder from the install tree to your desktop. Change the permission on the folder and all of its contents so they are not read only. Now you can change them as much as you want.

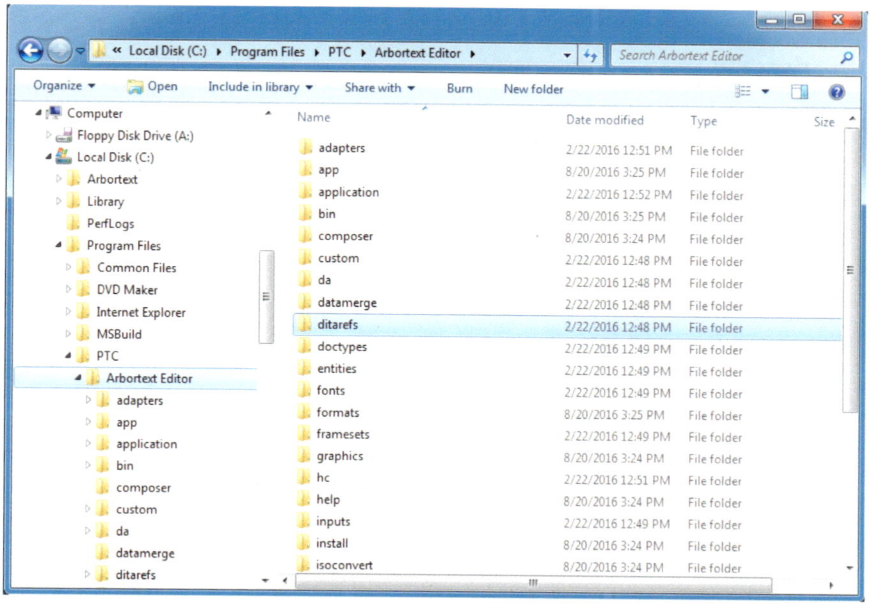

If you downloaded the sample files from the 102 website (http://102samples.single-sourcing.com), copy them into the **ditarefs** folder.

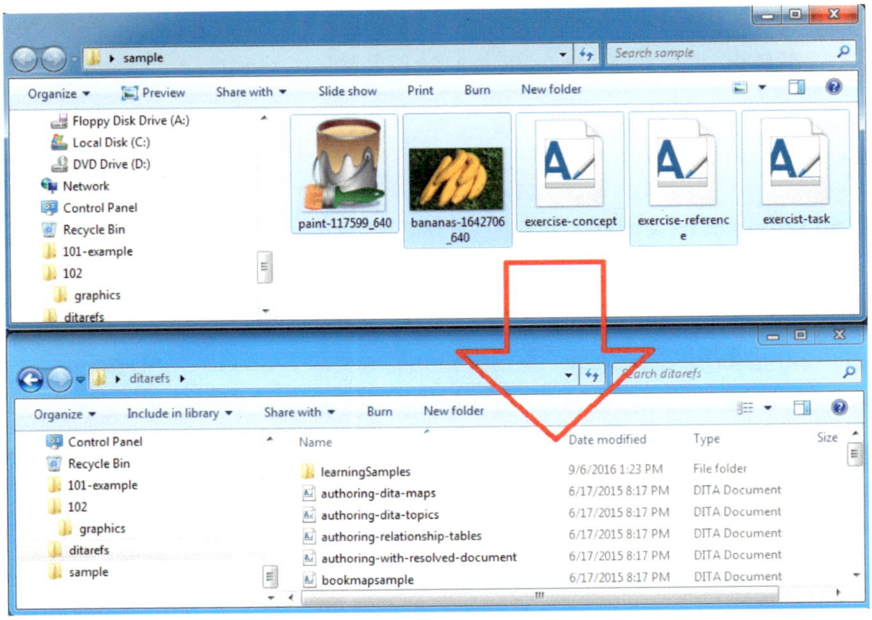

Open the **bookmapsample.ditamap** file.

Make these three changes to the bookmap:

1. Insert the three sample topics provided into the bookmap below the **DITA Authoring Overview (chapter)**:
   - exercise-concept.dita
   - exercise-task.dita
   - exercise-reference.dita
2. Move the **Authoring DITA Topics** topicref into the **frontmatter** and change its markup to **preface**.
3. Change the markup of the last chapter from **chapter** to **appendix**

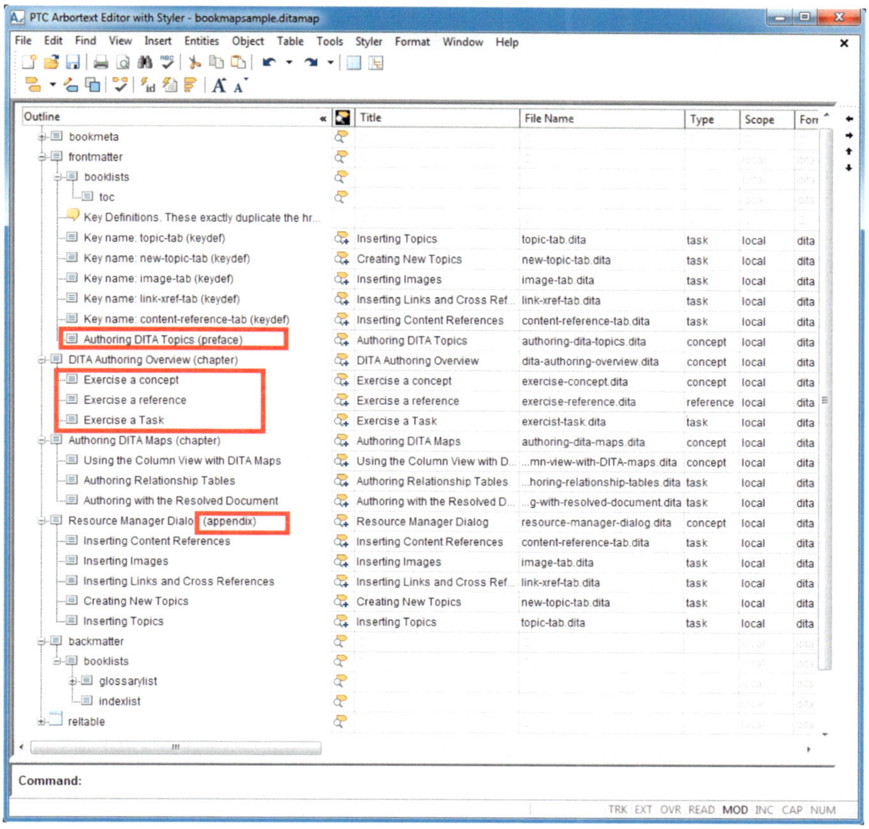

Now you have a strong variety of sample content to work with. You have a task, a concept, and a reference. You have more than one chapter, a preface, and an appendix. You have a TOC and an index. The sample topics that I provided have a variety of markup—both block elements and inline elements.

You can also create new sample files through the **New File** dialog box when you click the **Sample** radio button.

You can use either the OOTB sample with the sample I provided as a control document for testing your stylesheet. All the samples are valid, well formed, and always available.

However, none of the sample files represent *your content*.

You need to make sure that any sample document you create is typical of the content your organization creates now, or the way authors in your organization should create content once your stylesheet is ready.

For example, if you want to put the copyright date and your company name in your footer, you need to know where this information can be found so you can fetch the values from any source document. As long as these values exist in your source document *in a location that you can predict accurately*, you can use them in your stylesheet.

Create the stylesheet

Once you have a document open in the document editor, you can create a new stylesheet or open another Arbortext Styler stylesheet (a **.style** file). The only stylesheets that the Arbortext Styler stylesheet editor can manipulate are XML documents that conform to the **styler** document type provided by PTC.

When you begin work on a new stylesheet, it's a good practice to start with an empty stylesheet. Provide a completely clean starting point by creating a new empty stylesheet, then add the OOTB stylesheet as a module. Creating this modular structure gives you several advantages.

First, you get a clear line of demarcation between the OOTB stylesheet and your own formatting. You can show or hide any behavior or stylesheet element that you choose to override. The hierarchical nature of the way modules work means that you have a foolproof way to guarantee that your formatting has priority.

Second, you get everything included in the OOTB stylesheet by default in your stylesheet. All those default behaviors that are typical of technical publications content like typical page sizes, paragraph formatting, table handling, numbered and bulleted lists. Not to mention the special handling of formatting requirements like "**Table 4.** *Caption*" which has the word **Table** and the table number (both in bold) and the table title (in italics) and the complexity of cross-reference formatting. You don't have to code every element and every part of the stylesheet by yourself. You only have to make essential changes.

Third, because you have the OOTB as a module in your stylesheet, you can experiment as much as you want. You can always roll back experiments that don't work. The OOTB acts like a safety net.

And don't forget, if you swap out the old OOTB stylesheet with the new one when you upgrade to a new version of DITA, you get support for any new topics and elements for free.

It's convenient and it has everything to do with how modules work in Arbortext Styler.

When you create a new empty stylesheet, you will see that the tags in the document editor turn pink. A pink tag has no formatting associated with it. There are no rules in the upper half of the stylesheet editor window because the stylesheet is empty. All the tags are pink because there is no formatting information in an empty stylesheet.

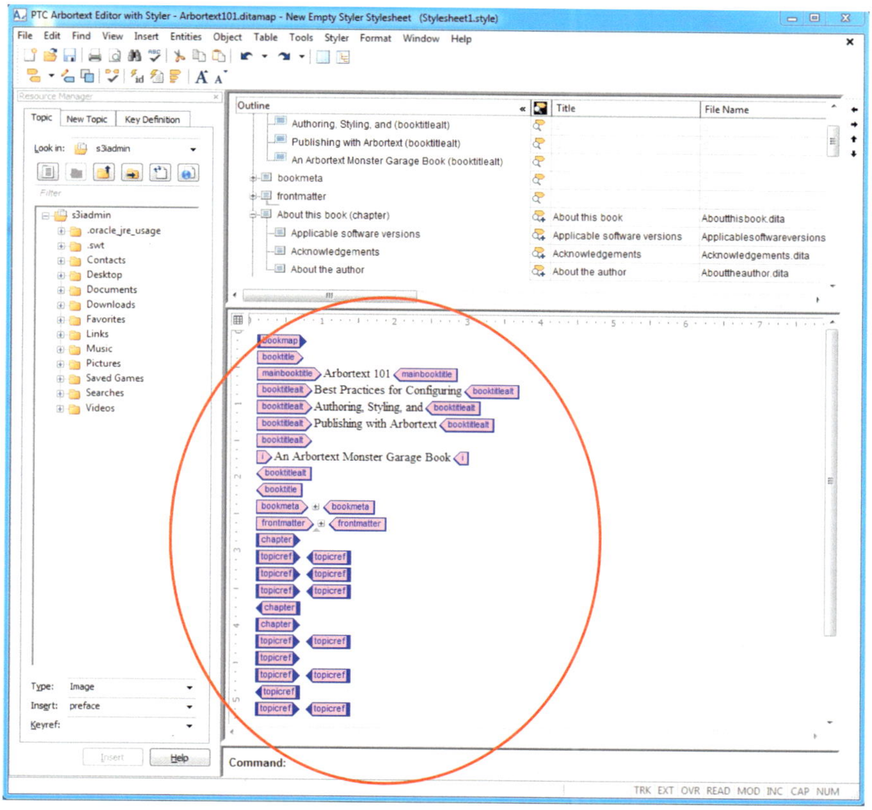

You can change that quickly by adding the default out-of-the-box stylesheet as a module or by adding a rule for formatting that element.

Expert tip!

We recommend always adding the default out-of-the-box stylesheet as a module in your stylesheet. Your stylesheet will go from empty to full simply by adding this module. You will get support for all elements in the doctype at one time and you can customize from there, overriding and changing the formatting as you go.

If you ever accidentally change one of the default stylesheet modules, you can simply copy them from the install tree to your **stylermodules** directory again.

Let's do it!

Make sure you have a sample document open.

From the **Styler** menu, choose **New Empty Stylesheet...**

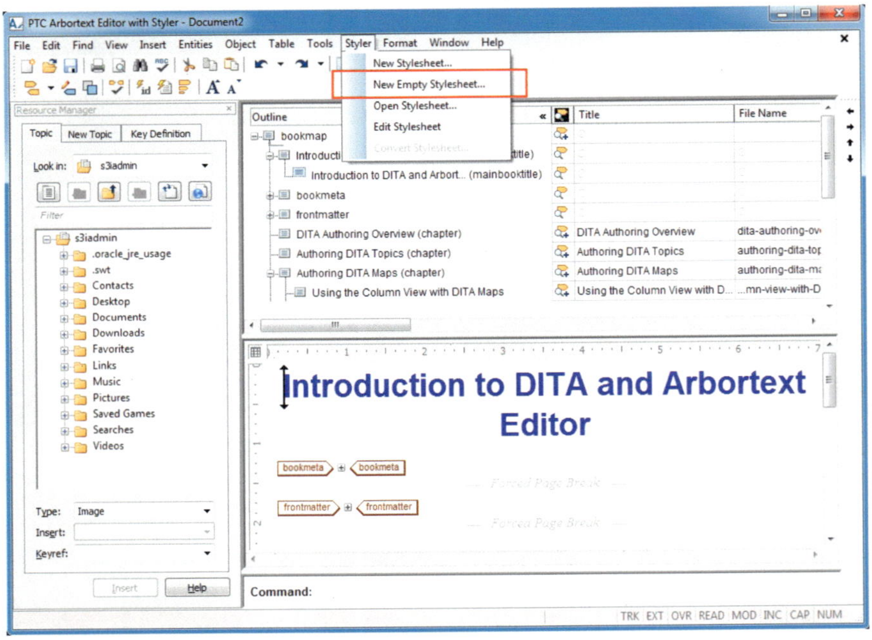

Give your stylesheet a **title** and **description**. You can look at the other tabs but do not change any of the other properties.

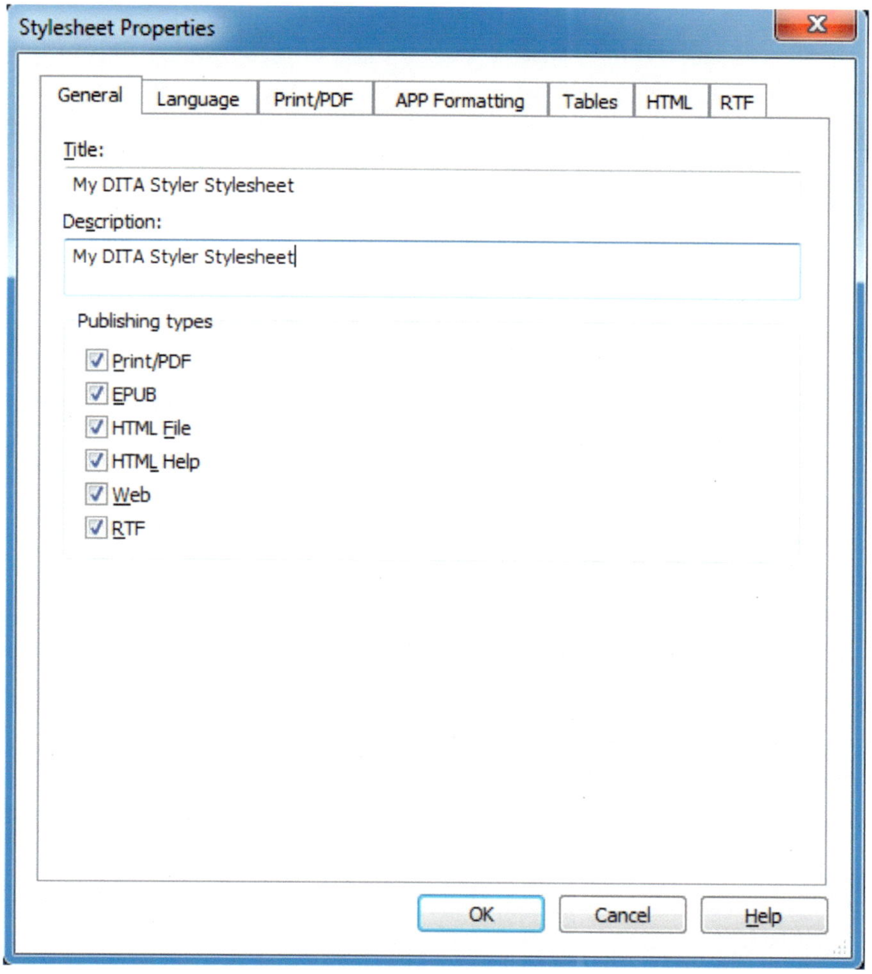

Click **OK** to close this window and open the stylesheet editor.

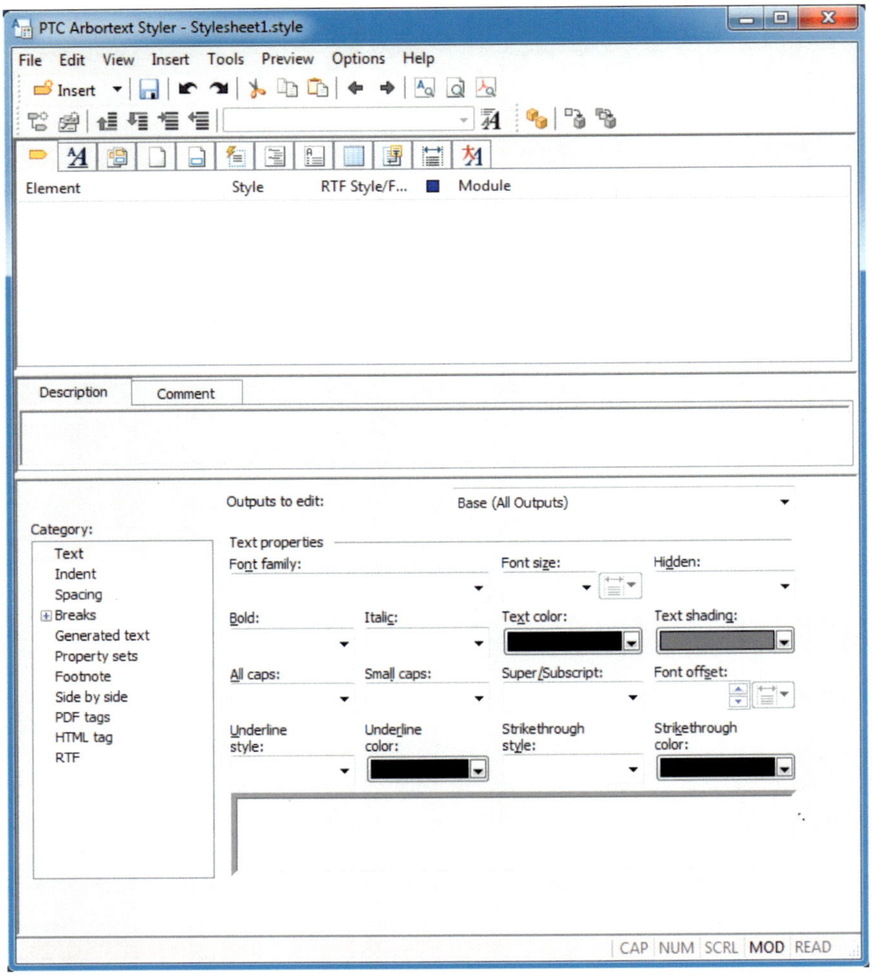

Click the **modules** button.

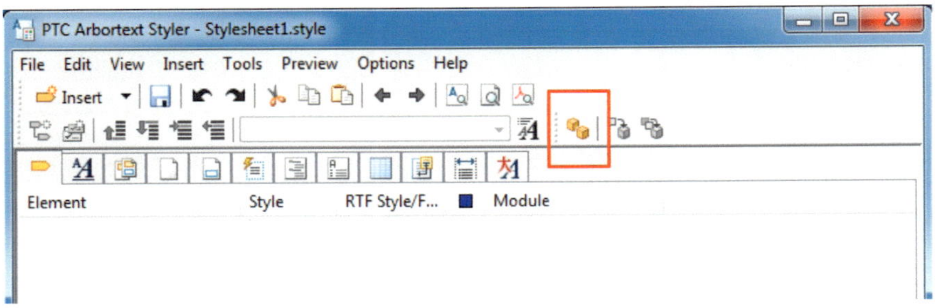

The **Modules** window opens.

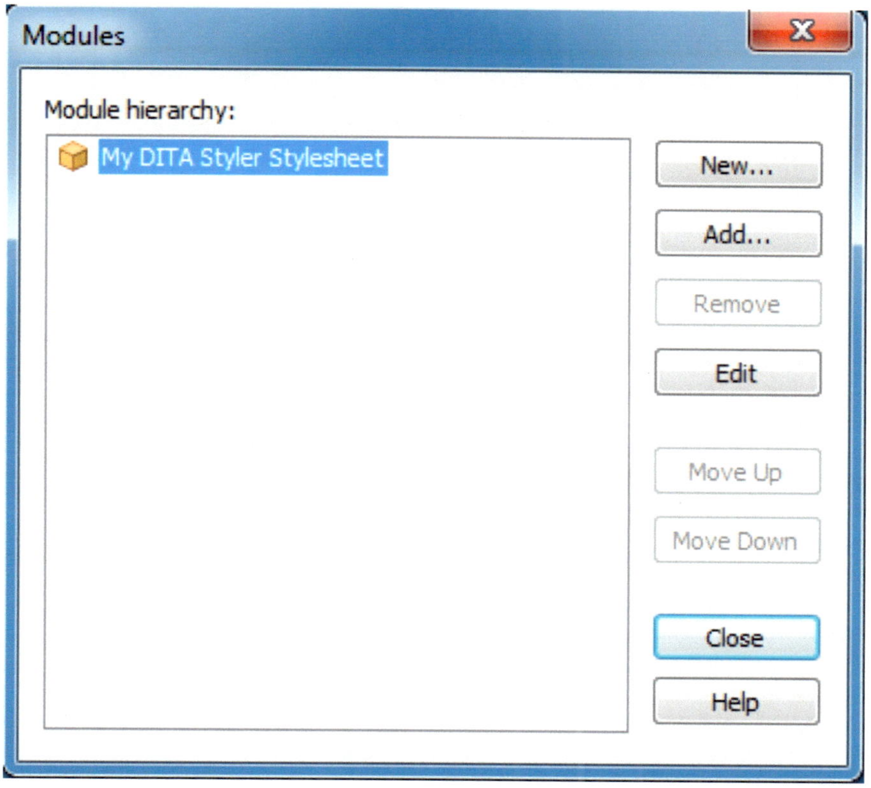

Click **Add**. The **Add Module** window opens and the **stylermodules** directory appears automatically.

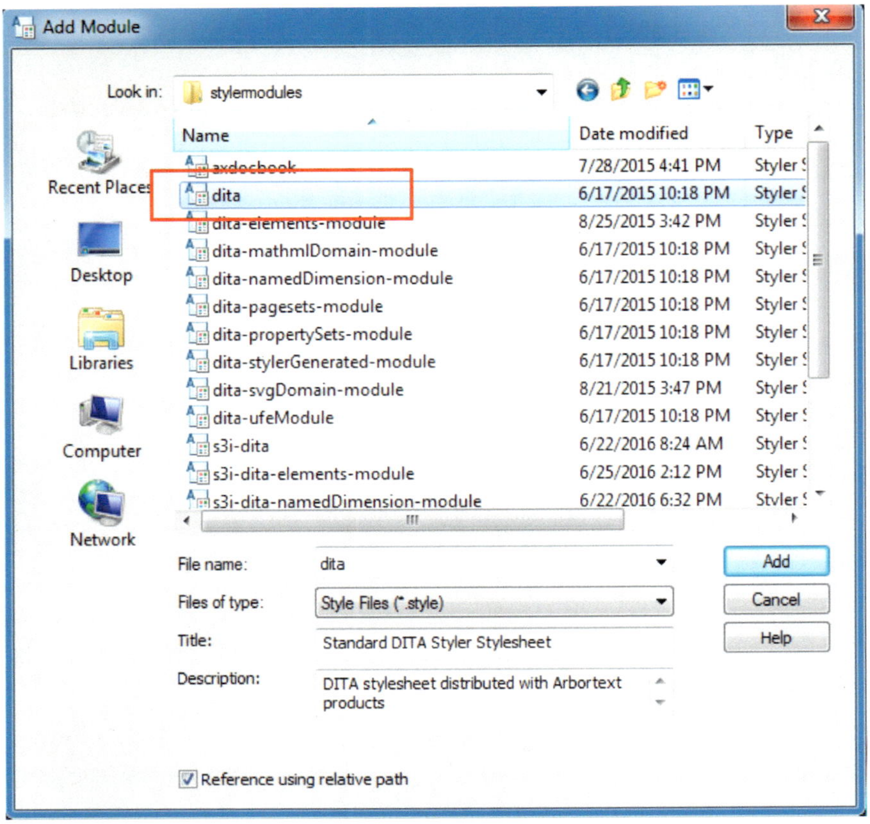

Select the **dita.style** file and click **Add**.

In the **Modules** window, you will see that the OOTB DITA stylesheet *and all of its subordinate modules* is a subordinate module in your stylesheet.

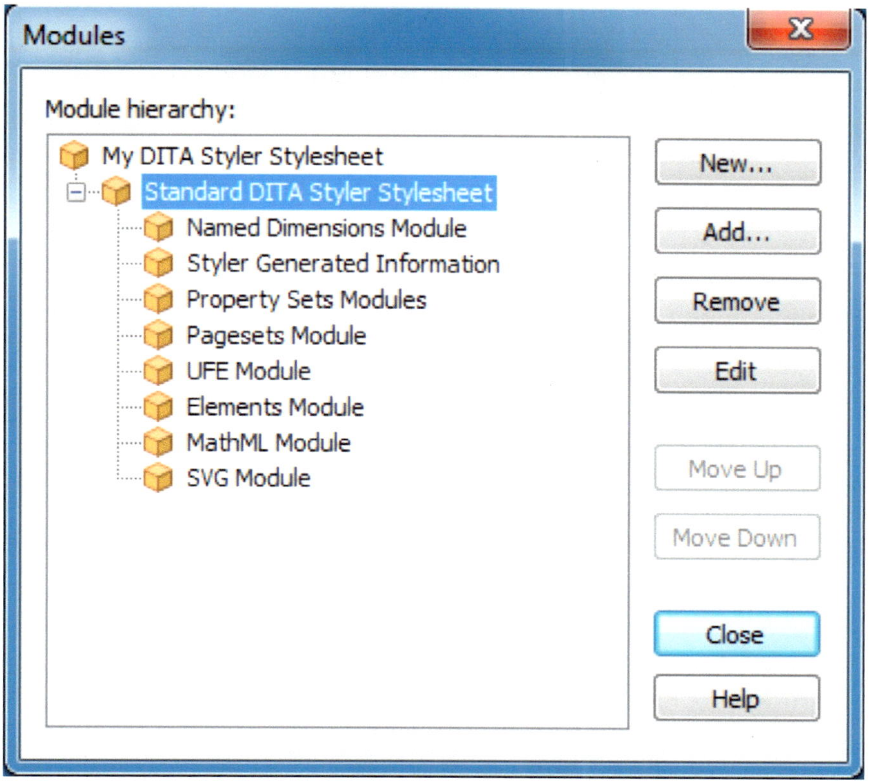

Close this window and return to the stylesheet editor. You will find that your stylesheet is no longer empty.

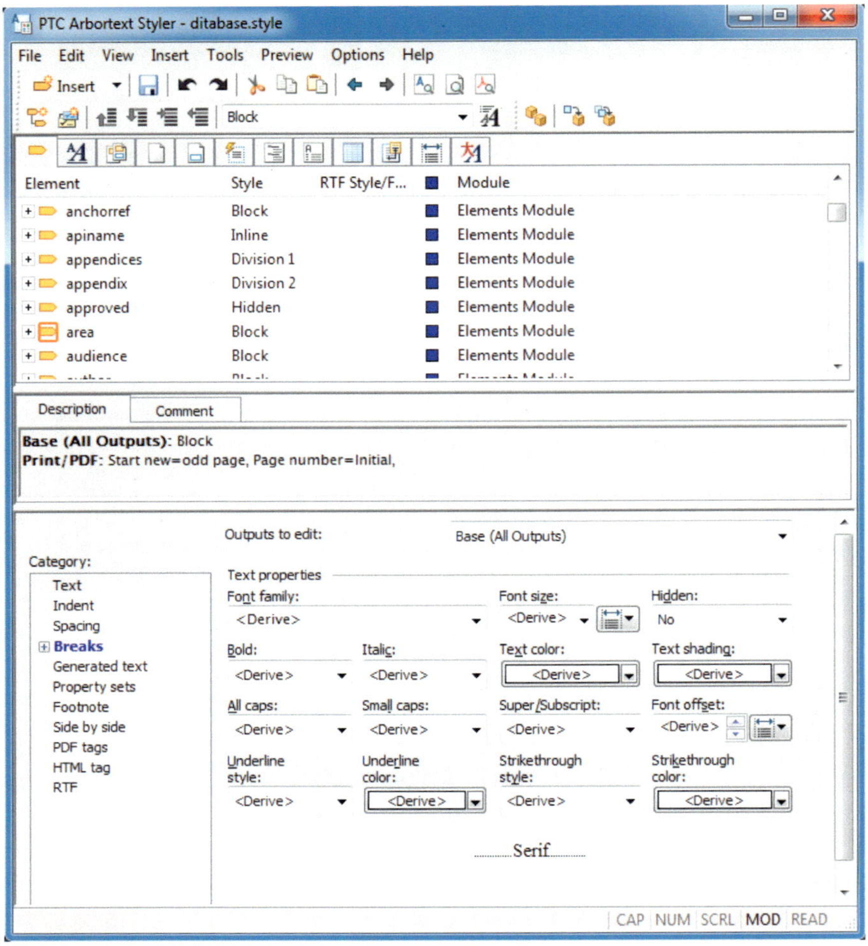

Save your stylesheet. Make sure you name it **ditabase.style** and save it in the **ditabase** doctype subdirectory in your custom directory.

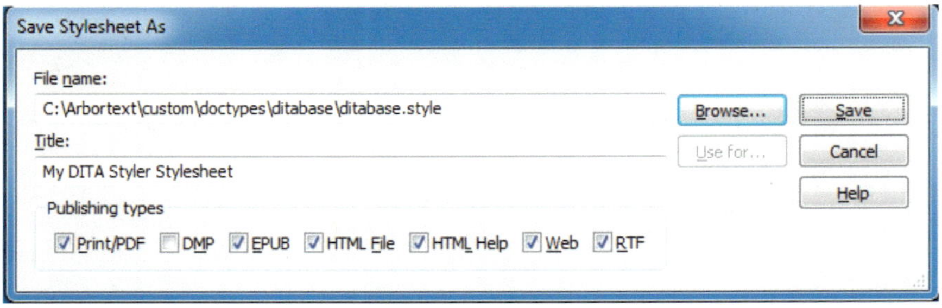

You can see the file in your custom directory.

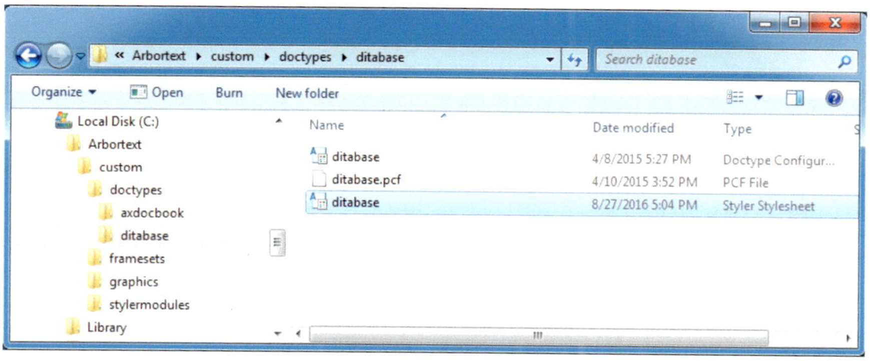

Close the style editor and close the document in the document editor.

That's it!

Chapter 4. Build a framework for print publishing

Topics Covered in this Chapter

♦ Create your own modules
♦ Understanding modules
♦ Set your default page dimensions
♦ Create your page sets
♦ Create new page types
♦ Create page regions
♦ Create generated content
♦ Create a cross-reference format
♦ Configure the page regions
♦ Configure page types
♦ Configure page sets
♦ Customize page layout (header and footer content)

This is the second step to understanding the power of Arbortext Styler. When you create a framework for print publishing, you understand how page sets, page types, page regions, and generated content work together. These four stylesheet components are the keys to understanding "putting content on pages."

I'm going to walk through a typical print layout. It provides the opportunity for me talk about everything needed to put content on a page. It's a pattern that you can replicate to create any kind of page layout including pages that are entirely generated like front and back covers, blank pages, and other unique layouts supported by the APP rendering engine.

APP can do advanced publishing tricks like bleed tabs, spilling text around a clipping path in a graphic, and dual-page spread layouts of tabular data. Styler is getting more powerful all the time adding more and more access to APP functionality.

It all starts with modules.

You should already have your custom directory set up and your first stylesheet created.

Open a sample document in Arbortext Styler, and let's get started.

Create your own modules

Create a **new empty stylesheet** and save it in the **stylermodules** directory.

Repeat the process until you have one new empty stylesheet for all of the following:

Filename	**Stylesheet name and description**
my-dita.style	My DITA Module
my-dita-elements-module.style	My DITA Elements
my-namedDimension-module.style	My Named Dimensions
my-pagesets-module.style	My Pagesets
my-propertySets-module.style	My Property Sets
my-stylerGenerated-module.style	My StylerGenerated
my-ufe-module.style	My UFEs

When you're done, you should have this:

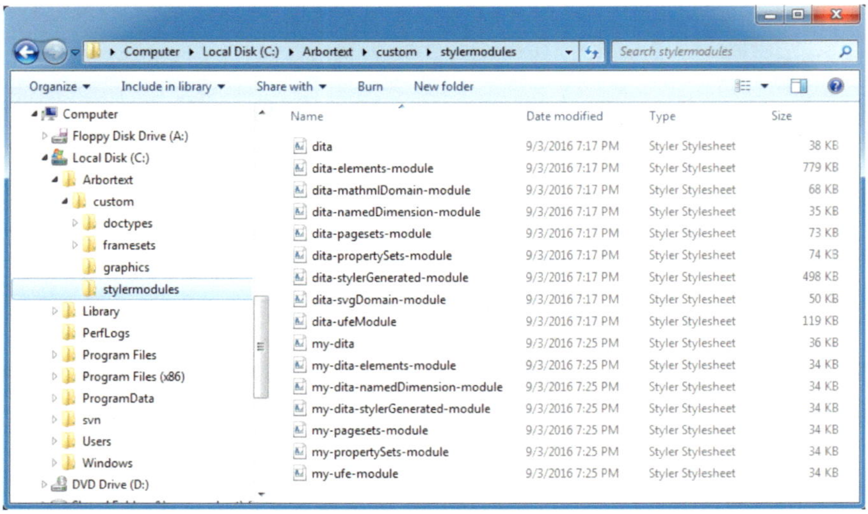

Add your new modules to your stylesheet. From the **Styler** menu, choose **Open Stylesheet...**

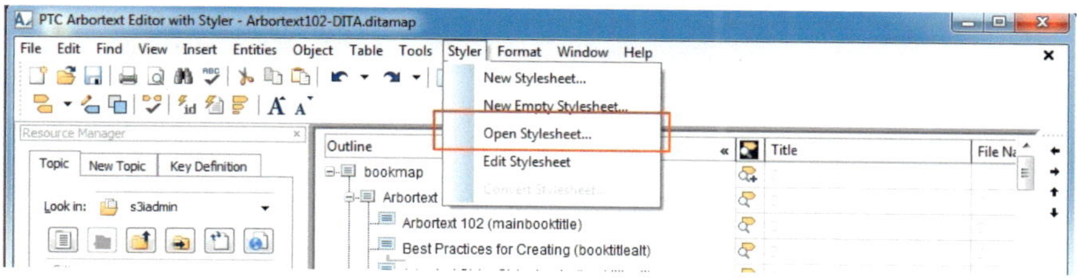

Choose **My DITA Styler Stylesheet**. You know that this is your stylesheet because can see the title matches what you specified, and the path is to your custom directory, as shown here:

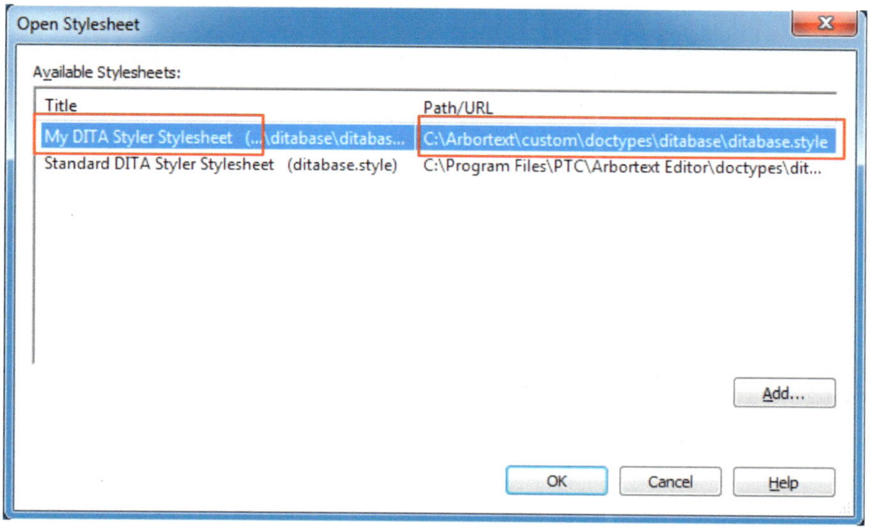

Once your stylesheet is open in Styler, click the **Modules** button to open the **Modules** window. Make sure that **My DITA Styler Stylesheet** is highlighted, then click **Add**.

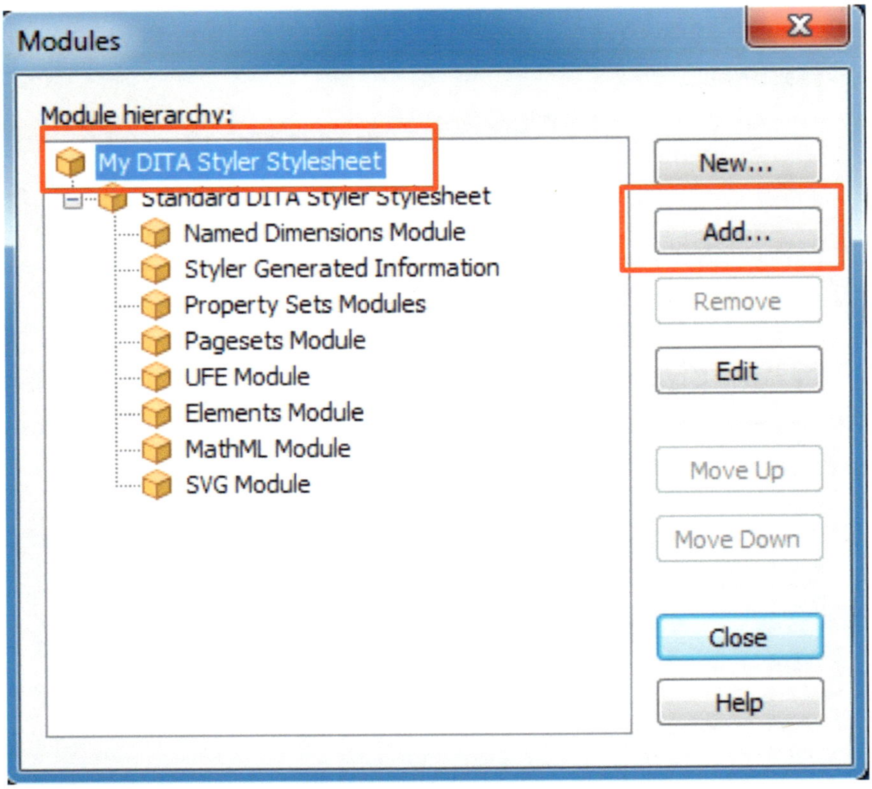

Navigate to the **stylermodules** directory in the file chooser window.

Select **my-dita.style** and click **Add**.

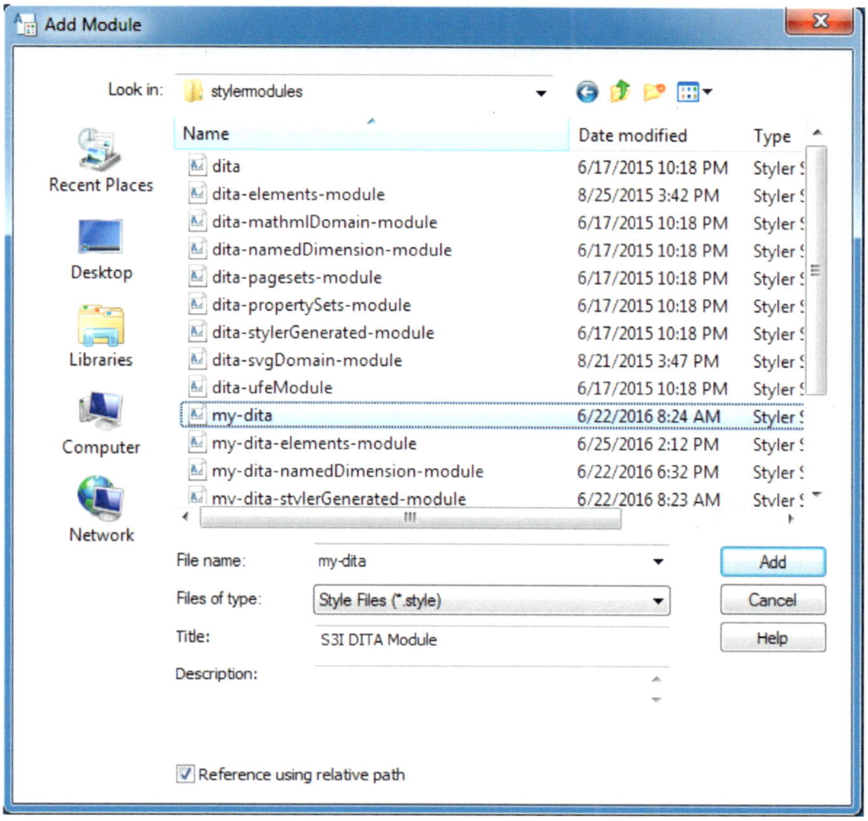

By default, your module will be added to the bottom of the hierarchy. Hierarchy is everything for stylesheet priority. You want your modules to have higher priority than the out-of-the-box stylesheets. That means that your modules need to be higher in the hierarchy.

To move the **My DITA Module** up in the hierarchy, select it and click **Move Up**.

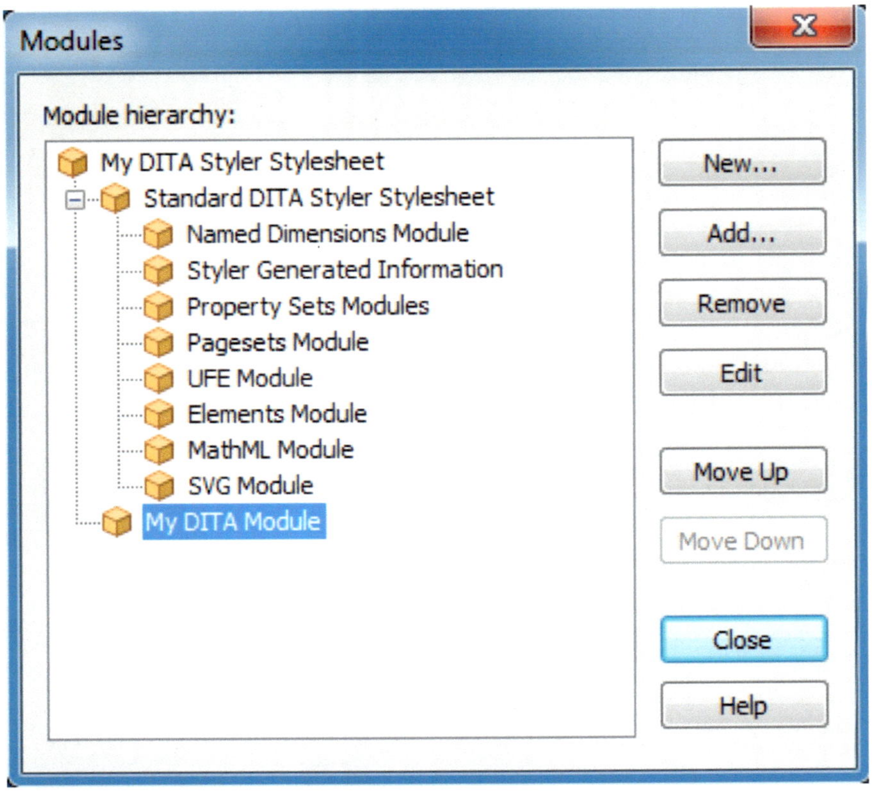

Your module now has higher priority than the **Standard DITA Styler Stylesheet**.

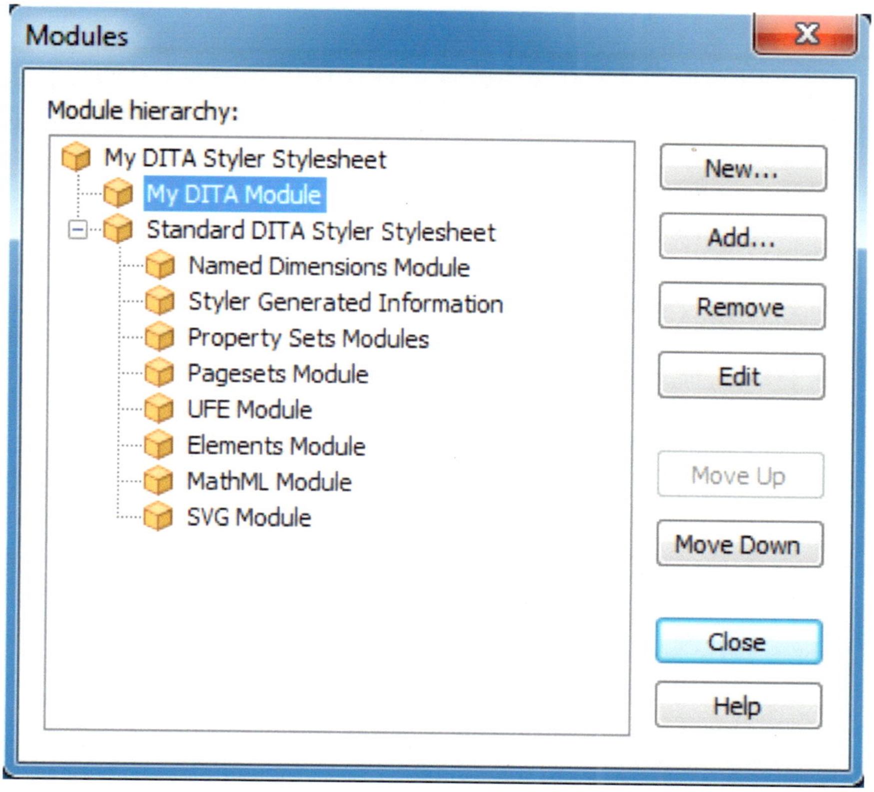

Add the rest of your modules in this order:

1. My DITA Elements
2. My Named Dimensions
3. My Styler Generated Module
4. My Property Sets Module
5. My Pagesets Module
6. My UFE Module

To guarantee that you create the module structure you want, make sure that you select **My DITA Module** *before* you click **Add**. This way, the modules will be added as subordinates to **My DITA Module**, which already has priority over the **Standard DITA Styler Stylesheet**. They inherit the priority of the module in which they are included.

Your final module structure should look like this:

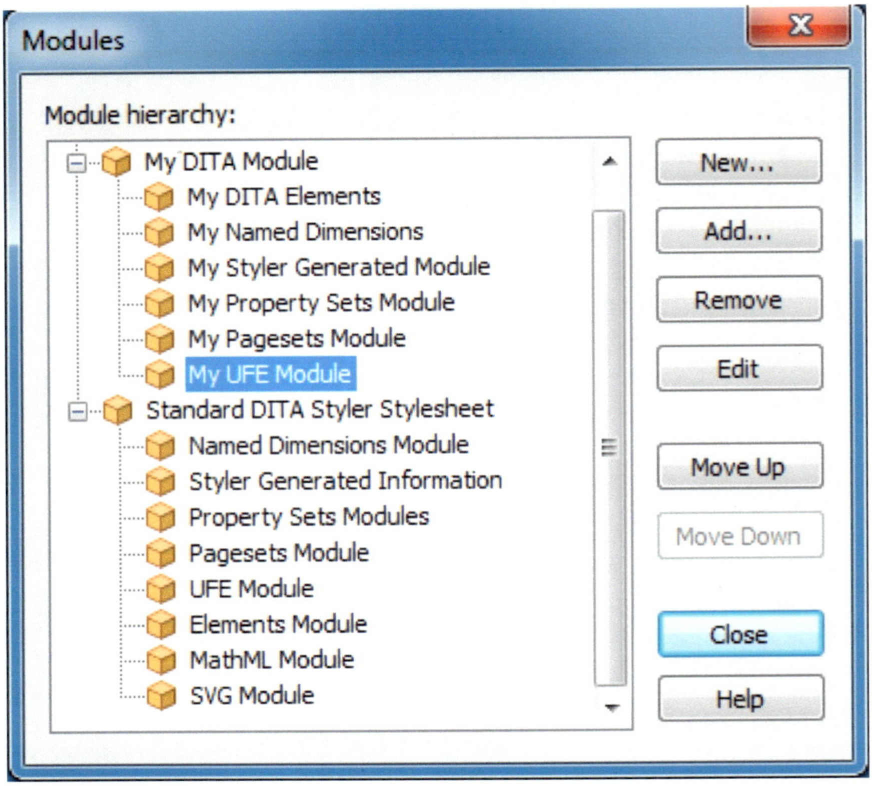

You now have all of the essential modules created and inserted properly into the module hierarchy. Save your stylesheet.

Now you can start really customizing your stylesheet!

Expert tip!

Notice that I had you create duplicates of the modules in the OOTB stylesheet. This is the best set of reusable components for future stylesheets. Page sets have wider usage when they don't depend on a doctype. Property sets are reusable units that define settings that aren't doctype dependent—like font face, font size, spacing, and indents. By encapsulating like components you increase your ability to reuse these modules in future stylesheets.

Understanding modules

Modules are the secret to Arbortext Styler. They do three things: They help you organize stylesheet components into logical units; they provide a way to guarantee hierarchical precedence for your customizations; and they encapsulate your customizations so you can reuse them in other stylesheets, reducing your workload and making it quick and easy to create a new stylesheet any time you need one.

One of the hardest adjustments for users coming from desktop publishing applications like Word is the freedom to change styling and layout at any time. When you go to XML publishing, you trade that for stability and consistency. Stylesheets take up the mantle of layout. Creating a stylesheet using traditional methods is typically a lengthy and costly exercise.

With Arbortext Styler, any stylesheet can be a module of another stylesheet. This gives you an unprecedented leg up if you must create a new layout. Just like you included the OOTB stylesheet as a module in your first stylesheet, you can include that stylesheet in another one.

You included the OOTB stylesheet as a module in your stylesheet. You added a complete library of formatting rules, page layout rules, generated content rules, and other rules to your stylesheet. You can use them as they are or, because you also added your own modules, you can override them.

No stylesheet can have two rules with the same name and the same precedence. When two rules have the same name but different precedence, we say that one rule **overrides** the other.

To override a rule, you duplicate that rule and put the duplicate in your module. By duplicating the rule, you are establishing the override. You keep the same name but change the precedence. By putting the duplicate rule in your module, which is already higher in the module hierarchy, you are guaranteeing the precedence of your rule. You are effectively substituting your rule for the rule provided out of the box. You can freely duplicate and add rules to your modules with the confidence of knowing that you are ensuring the precedence already established in your stylesheet module hierarchy.

The Arbortext Styler UI helps you see rule precedence. Modules serve to assist in that regard. Here is an example:

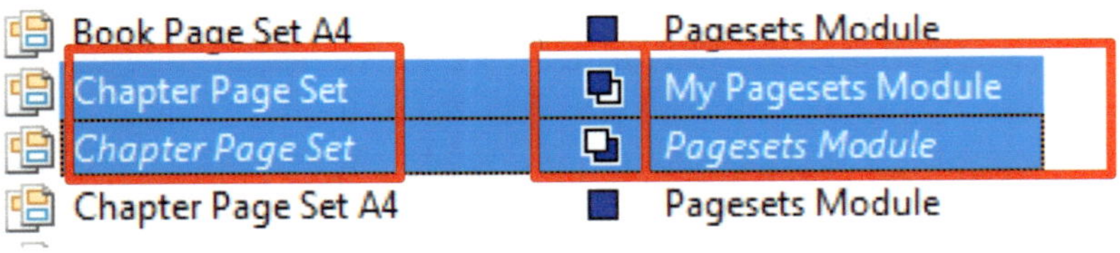

Both of these rules are named **Chapter Page Set**. The UI shows the module name for each rule, so you know that they have different priority levels. You can check the module hierarchy window, or you can use the other clues in the UI.

First, notice that one rule is italicized and the other is not. When a rule is italicized, it is being overridden; it is not the active rule. A rule that is not italicized is active. Second, notice the blue box icons. A blue box means that a rule is active. A blue box over a white box means that the rule has the highest precedence and is overriding another rule. A white box over a blue box means that the rule is inactive because it has lower precedence than another (active) rule.

The other advantage to seeing the module name for each rule is that you can group rules together visually by module name. To do so, click the **Modules** column header in the upper panel:

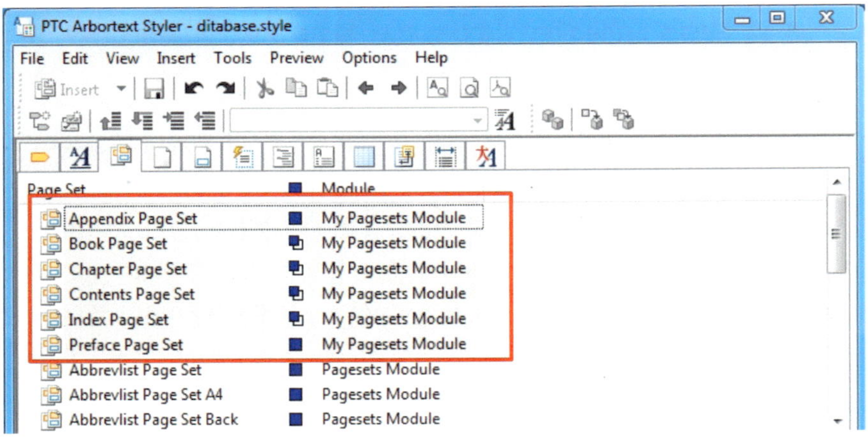

Modules help you logically separate rules by module name. You can easily identify which rules are and are not in your module by looking at the **Modules** column. When you find the rule you're interested in, you can resort by rule name and see what other rules are in the stylesheet, in what other modules, and at what other precedence levels.

Set your default page dimensions

The **Sizes** tab is where you define your page dimensions. Dimensions are used throughout in the stylesheet to determine how much space can be used for regions like headers, footers, margins, and page body.

In the out-of-the-box stylesheet, the inside mirror margin is .25in bigger than the outside mirror margin. When the inside mirror margin is greater than the outside mirror margin, the book has a gutter. A gutter is the extra space required by the book binding.

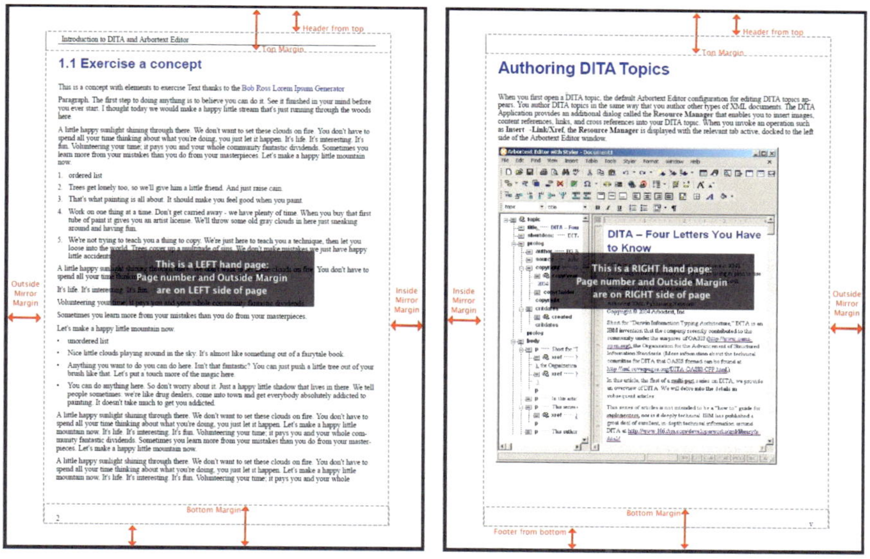

If you don't publish your books as PDFs instead of printing them, you probably don't want to have a gutter. I'll show you how to change the sizes of the margins so that all margins (top, bottom, left, and right) are the same.

Let's do it!

Click the **Sizes** tab.

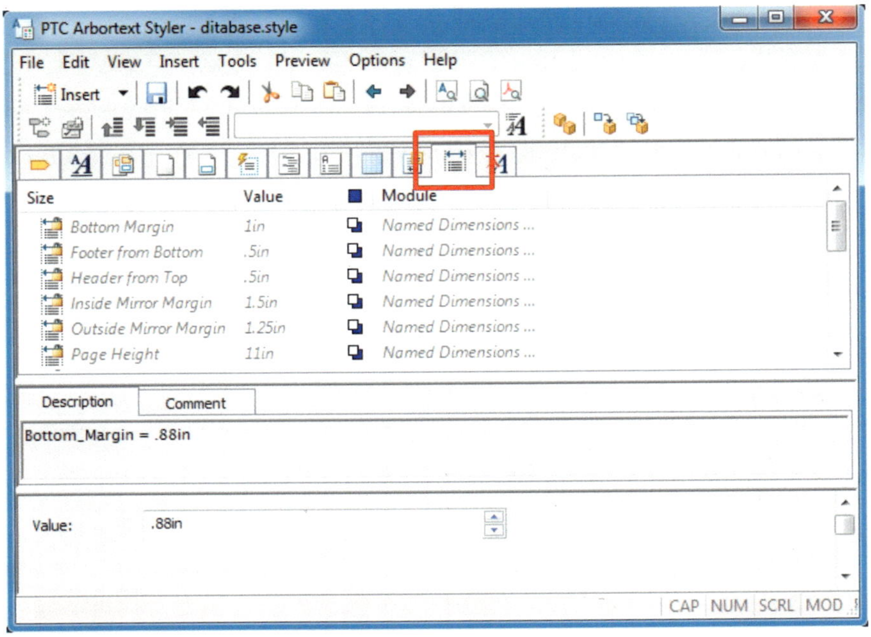

Select **Inside Mirror Margin**.

Click the **Copy to Module** button  to open the **Copy to Module** window. Select the **My Named Dimensions** module and click **OK**.

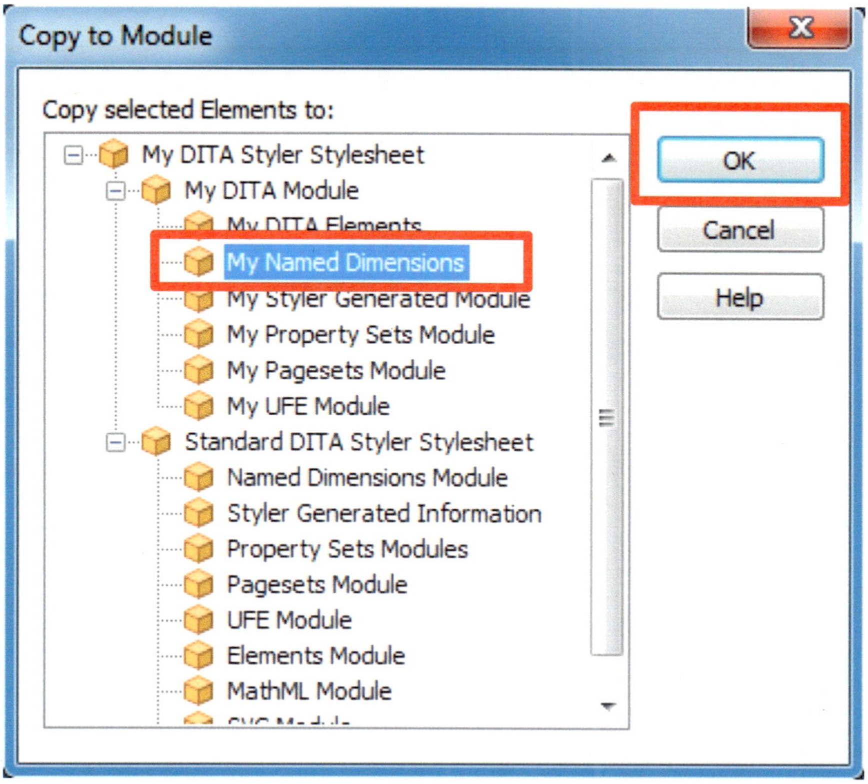

The Inside Mirror Margin rule now has **My Named Dimensions** as its module name**.** Select the **Inner Mirror Margin** rule and change the value in the lower pane from 1.5in to 1in.

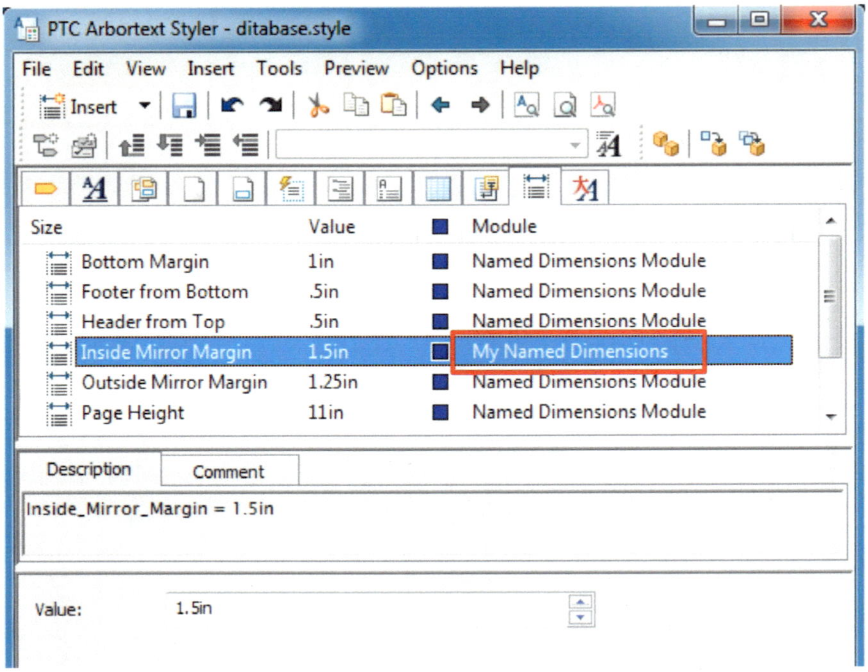

Repeat this process for the **Outer Mirror Margin**: Copy the rule to your module and change the value.

When you're done, you should have this:

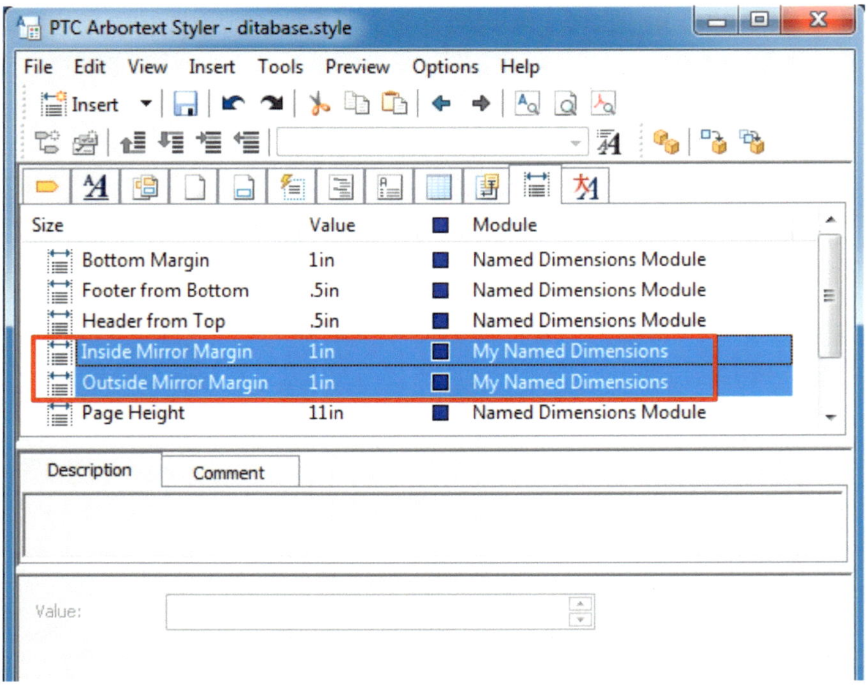

Both rules have a blue box. Neither rule is italicized. Your rules have precedence. The next time you publish using this stylesheet, the inner and outer margins will have the values you set (1in).

Once you're happy with your page dimensions, you can create the page sets. They will automatically use the margins you set here.

Save your stylesheet.

> **Expert tip!**
>
> Although the image above does not show that two sets of rules (the pair in **My Named Dimensions Module** and the pair in **Named Dimensions Module**) exist in the list of rules shown in the upper panel of the **Sizes** tab, both sets of rules (one italicized, one not) are in the stylesheet. They are just not shown in this picture. In order to see this exact image in your Styler window, go to **View→Show Duplicate Definitions** and make sure it is **unchecked**.

Create your page sets

Arbortext Styler's default DITA modules come with several page preconfigured page sets. You have access to all of them because you included the OOTB stylesheet as a module in your stylesheet.

We recommend making some changes to make things easier for new stylesheet designers. The OOTB stylesheet can serve as a demonstration piece as well as a starting point. As a result, it can be complicated to decipher. To prevent you from having to figure out things over and over again, I'll show you how to configure your page sets the way we teach our customers.

We recommend configuring one page set for each major section of the book—preface, table of contents, chapters, appendixes, and index. All of these parts differ slightly from each other. Some have Roman numerals for page numbers; some do not. Some have static words in the header; some have running heads.

To make things easy, we recommend configuring all of these page sets so that they work correctly for each section of your book. While not the optimum configuration from a reuse perspective, this method can help new stylesheet developers avoid unexpected formatting problems (for example missing or incorrect header text). This configuration also makes modifications easier down the road—especially when you are pressed for time.

Our approach uses consistent naming practices to help you organize the components into logical groups that are easy to follow.

You want all of the following page sets:

- Book Page Set
- Chapter Page Set
- Contents Page Set
- Index Page Set
- Preface Page Set
- Appendix Page Set

All but the last two (Preface and Appendix) are already in the OOTB stylesheet. You are going to create overrides for them and create two new page sets.

You want to create overrides just like you did with the inner and outer margin rules.

> **Expert tip!**
>
> You never want to change the OOTB rules. Instead you always want to *override* them by creating copies of them in your own modules. Creating copies of the existing page sets in your own module gives rules precedence and preserves the changes you make in your modules.
>
> To guarantee that your changes are safe, always create a duplicate and override. That way, your rules won't be affected by any changes that PTC makes to the OOTB stylesheets in the future. Your rules will always have priority and they will only be in modules that you control.

Let's Do It!

Click the **Page Set** tab. Right now, the only page sets are those that come with the OOTB stylesheet.

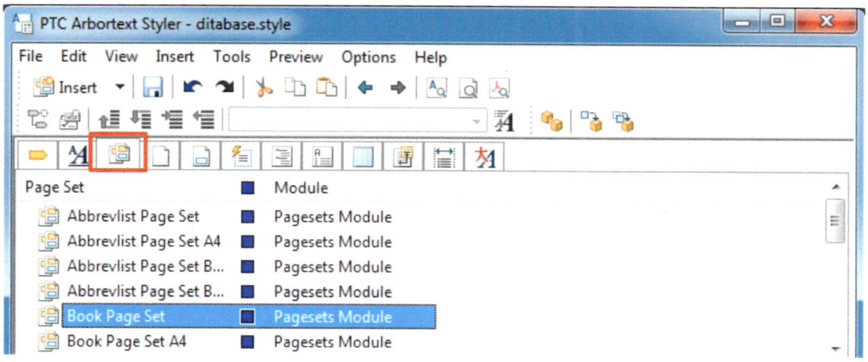

Select all the page sets you want to duplicate (Book, Chapter, Contents, and Index) and click the **Copy to Module** button.

In the **Copy to Module** window, select **My Pageset Module**, and click **OK**.

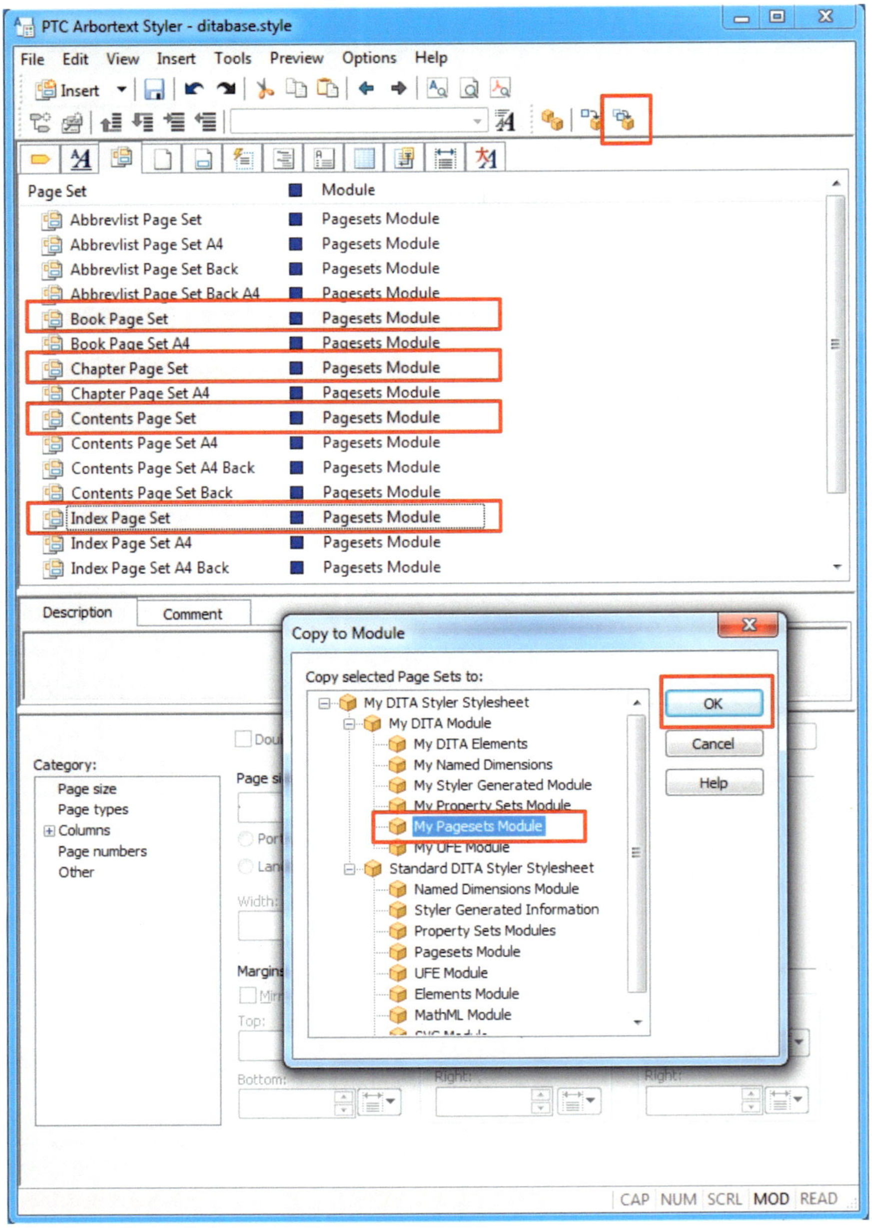

Next, you'll need to create the two missing page sets (Preface and Appendix). To create them, perform the following steps.

First, select **Chapter Page Set**. Then click **Copy** followed immediately by **Paste**.

You'll see a warning message about precedence.

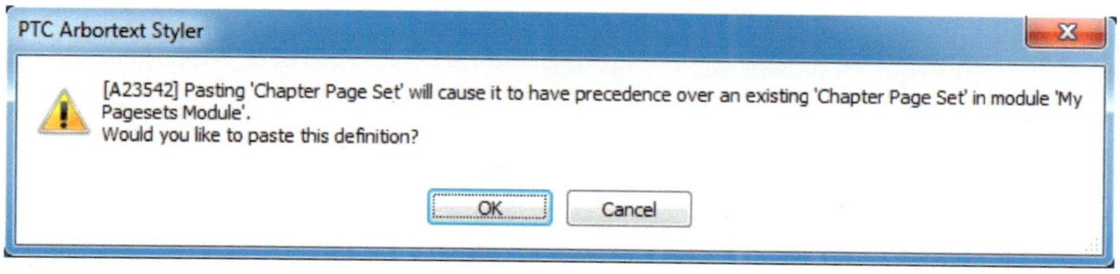

You can ignore this message for now because your paste action is temporary. You'll change the name of the pasted page set, which will eliminate the precedence issue. You're simply copying and pasting as a short cut.

Click **OK** to dismiss the warning. You'll get a new copy of the chapter page set that doesn't belong to any module. After you paste, you can edit the name.

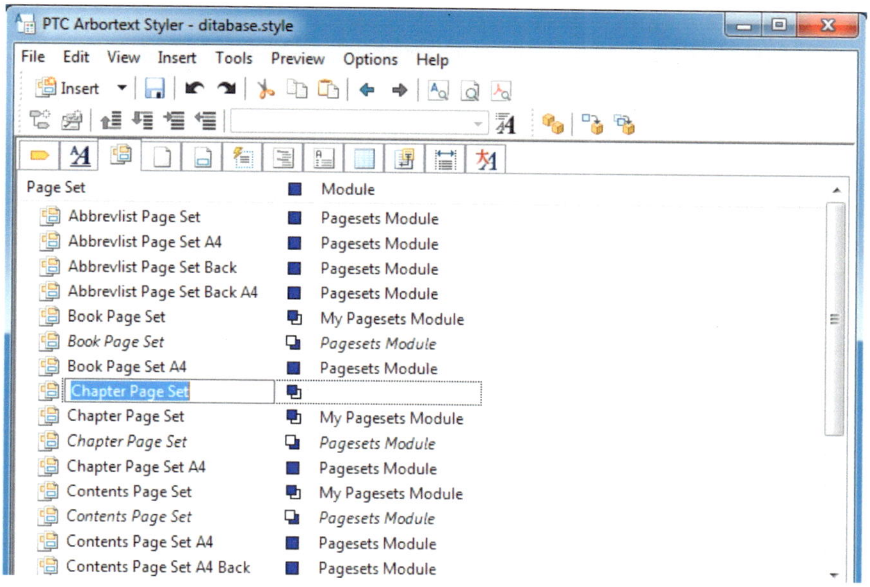

Change the name to **Appendix Page Set** then **move** it to your page set module.

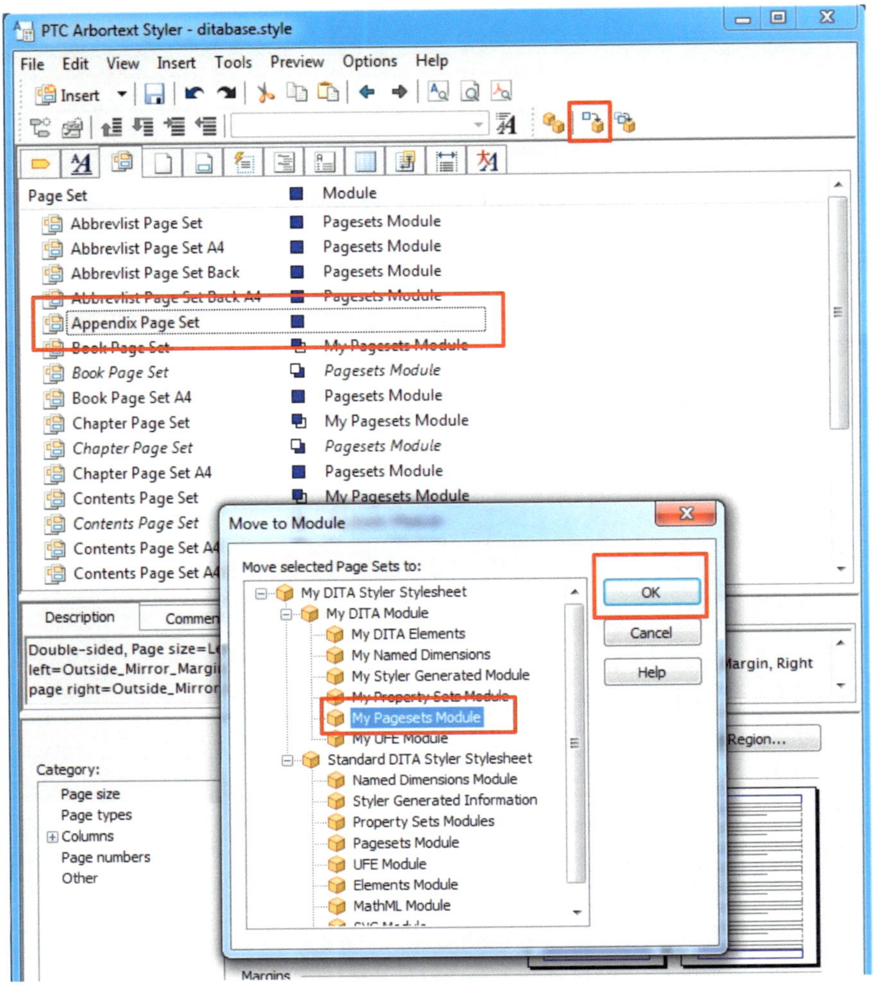

Create the preface page using the same method, but use the **Contents Page Set** instead because a preface page set is more similar to a contents page set than to a chapter page set.

Select the **Contents Page Set**, copy, paste, change the name to **Preface Page Set** and move it to your page sets module.

When you're done creating all the page sets, you should have this:

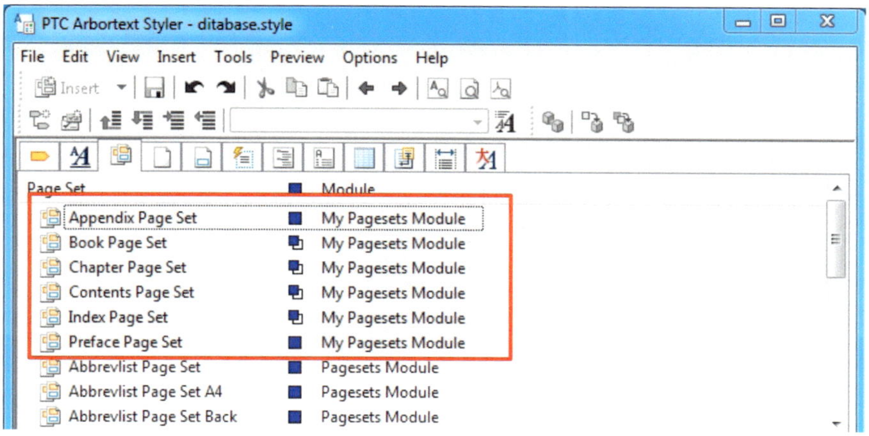

Save your stylesheet.

Expert tip!

In order for your Styler window to match the picture shown, sort by module by clicking on the **Module** title. This will group the rules into modules and your view will match.

Examine a page set

Now that you have all the page sets created, take a closer look at one of them.

Click **Chapter Page Set** in the upper panel, then the **Page types** category in the lower pane. Look at the different page types available in a page set.

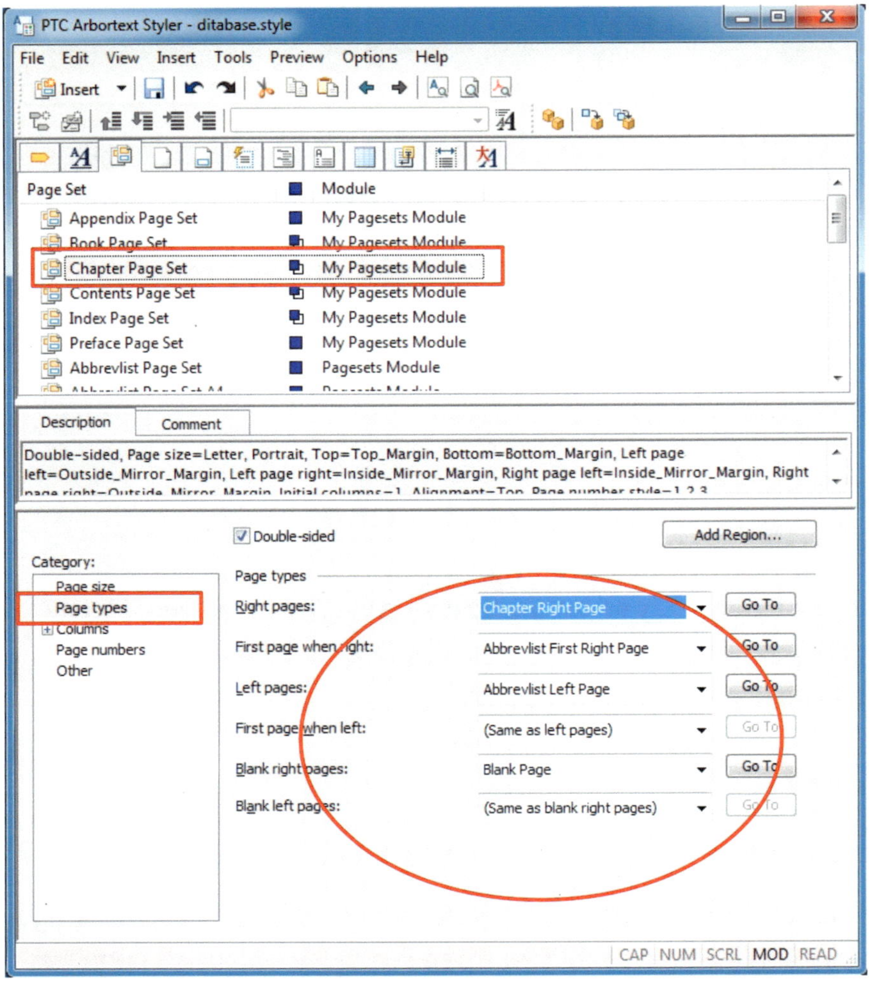

Notice that you have a mix of mismatched page names. The page set is using **Chapter Right Page** for right pages, but it is using **Abbrevlist First Right Page** for the first right page in a chapter. It's using the **Abbrevlist Left Page** for left pages.

Click the **Contents Page Set** in the upper panel, and compare it with what you saw in the **Chapter Page Set**. **Contents Page Set** looks a lot like chapter, but it has a **First Left Page** configured.

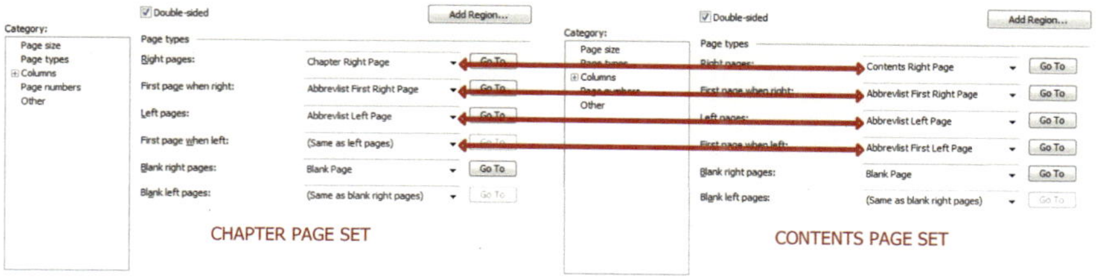

Take a look at the other page sets and you will see that the different page types aren't named or configured consistently. You're going to change that. Before you do so, though, you need to create some new page types. Once you've done that, you can return and reconfigure the page sets.

Page sets and page types

Page sets define collections of pages used to control layout. A page set includes settings for page configuration, like number of columns, how page sizes and dimensions are applied, page number format, and how page types are organized. A page set can specify the page types that should be used and the conditions for their use.

For example, a chapter starts on a new page, but not just any new page. A new chapter starts on a new *right* page. The first page in a chapter may differ from the other right pages in the chapter—for example, the first page may not have header content. These different conditions mean that you need different page configurations. The page set defines the situation and maps page types to those situations.

You can configure a page type for each of the following situations:

1. Right page
2. Left page
3. First right page
4. First left page
5. Blank right page
6. Blank left page

Page sets organize page types. However, you can add existing page types to a page set. When you looked closely at the chapter page set, you saw an inconsistently named page type. Before you can finish configuring the page sets you created, you must create the necessary page sets.

Expert tip!

Don't forget that you can sort the upper pane by module name by clicking the **Module** column head. Doing so can provide a quick and easy way to see whether you have any stray rules that are not assigned to a module. It can also give you a convenient view of all of your work.

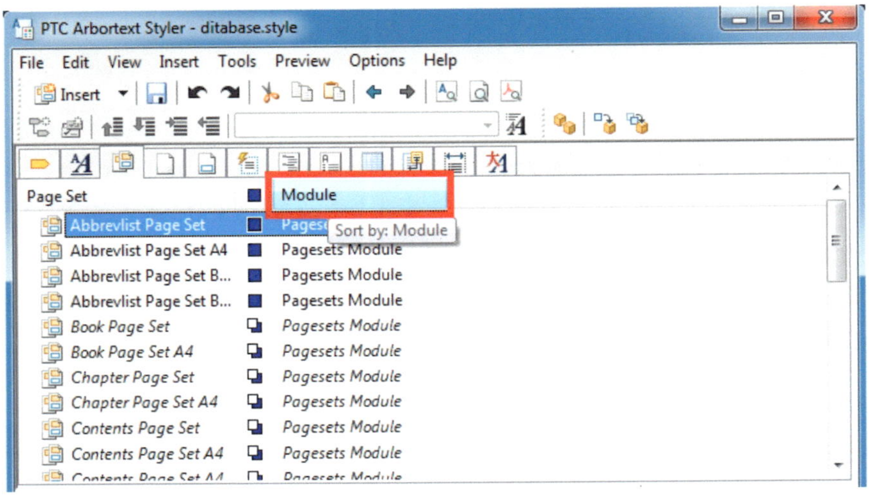

Create new page types

As with page sets, including the OOTB stylesheet as a module means that you have access to several preconfigured page types. And, as with page sets, we recommend creating page types that follow consistent naming practices.

Click the **Page Type** tab. You can see that not many page types are included in the OOTB stylesheet. It is missing many page sets that you want in order to configure all the page sets you created.

Because you created a lot of page sets, you're also going to create a lot of page types. You want all the necessary page types for the page sets you created. To create those page types, use the same copy-paste-rename-move method to create the page types you used to create your page sets.

Let's do it!

Click the **Page Types** tab.

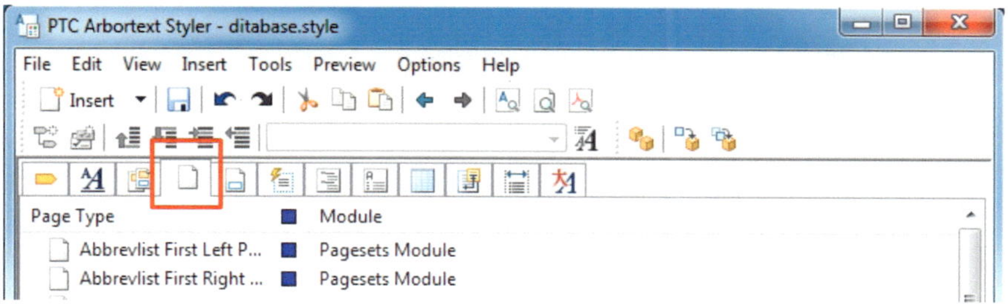

Copy these page types to your page sets module:

- **Chapter Right Page**
- **Contents Right Page**

Copy **Index Page Set right** to your page sets module and rename it to **Index Right Page**. Ignore any warnings about precedence. A rule can only override another rule with the same name. When you change the name, the rules no longer match and the warning no longer applies.

Create the following page sets by copying an existing page type and then pasting, renaming, and moving it to your page sets module. Again, ignore any warnings about precedence.

Copy and paste the **Abbrevlist *Right* Page** to create these page types:

- Appendix Right Page
- Preface Right Page

Copy and paste the **Abbrevlist *First Right* Page** to create these page types:

- Chapter First Right Page
- Contents First Right Page
- Index First Right Page
- Appendix First Right Page
- Preface First Right Page

Copy and paste the **Abbrevlist *Left* Page** to create these page types:

- Chapter Left Page

- Contents Left Page
- Index Left Page
- Appendix Left Page
- Preface Left Page

Move all of the page types you created to your page sets module.

When you're done you should have this:

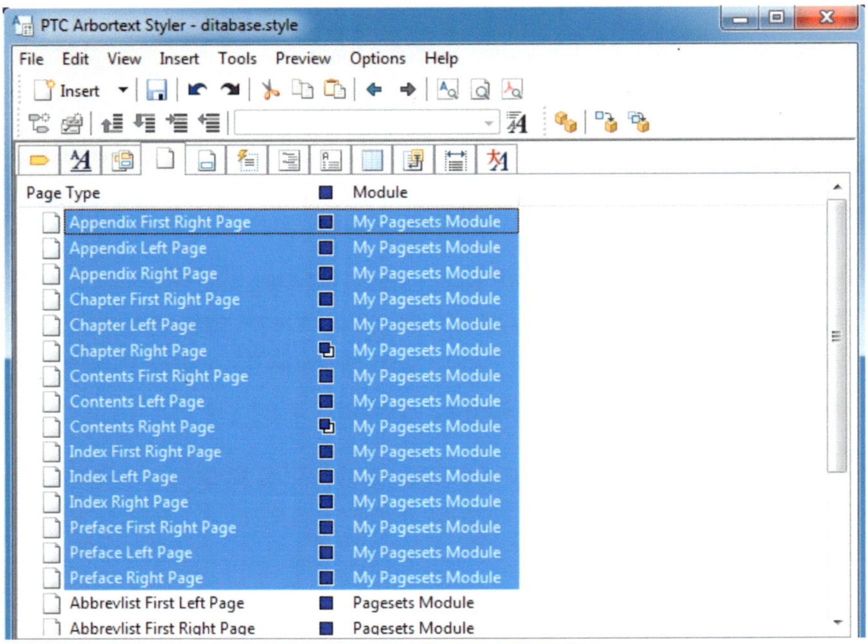

Save your stylesheet.

Expert tip!

Choose a right page to duplicate other right pages. Choose a left page to duplicate other left pages. Choose a first right page to duplicate other first right pages. Choose a first left page to duplicate other first left pages.

Examine a page type

Now that you've created your page types, take a closer look at one of them. Click **Chapter Right Page** in the upper panel, then look at the **Page regions** pane in the lower panel:

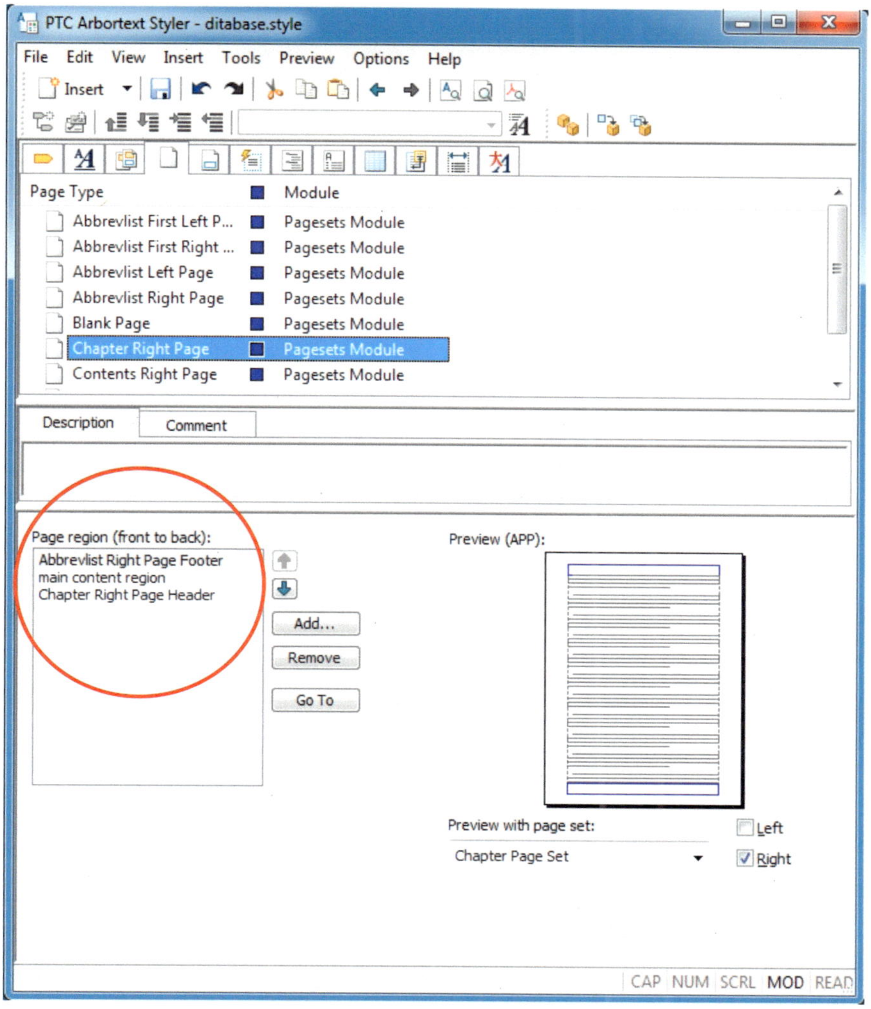

The inconsistent naming we saw in the page sets only gets more confusing in the page types. The **Chapter Right Page** uses the **Abbrevlist Right Page Footer** (page footer), the **main content region** (page body), and the **Chapter Right Page Header** (page header).

Take a look at the other page types, and you'll see that the different page regions that make up any given page type aren't named or configured consistently. You're going to change that. Before you can do so, you have to create some new pages regions because you can only add page regions that already exist. Once you've done that, you can return and reconfigure the page types. (Starting to sense a pattern?)

Expert tip!

Page sets, page types, page regions, and generated content have tightly woven dependencies. You can keep it all straight if you visualize working across the tabs left to right, and then back again right to left.

Page types and page regions

Every page type defines the regions that appear on it. These regions can include a header, a footer, a bleed tab, a watermark, or anything else. Page regions are like layers in a graphic file. Layers are applied to a page bottom to top (back to front). The bottom-most layer is the back. The top-most layer is the front.

Select **Contents Right Page** in the upper panel to take a closer look at the regions configured for this page type.

In the lower panel, you can see the list of page regions and the order of their application on the left side. On the right side, you can see a visualization of the page and its regions. If you click a region in the **Page regions** pane on the left, you'll see it highlighted in green in the **Preview** on the right.

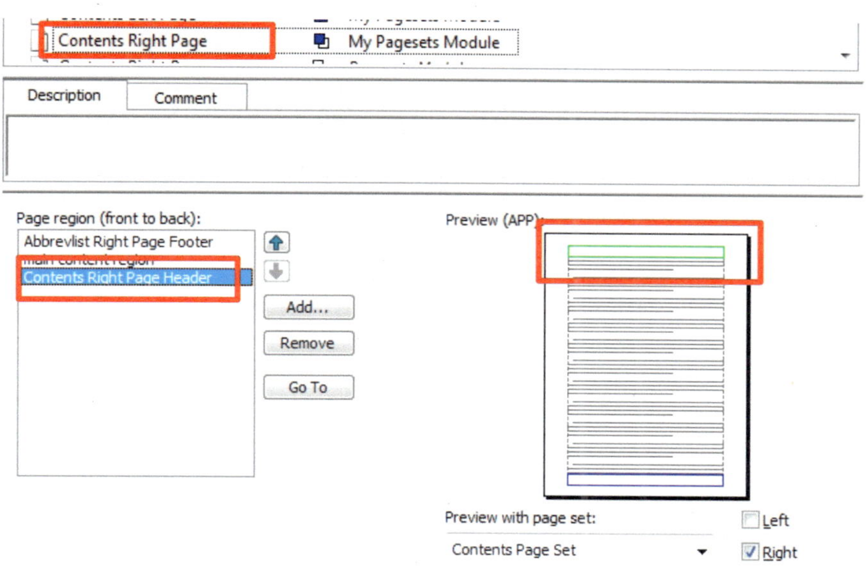

The **Preview** visualization can show you where page regions are located and how big they are. You can see how they overlap, if they do. It's a helpful tool especially when you're working with complicated page types.

Expert tip!

The Arbortext Styler UI provides shortcuts to the items that connect to each other, like page set and page type or page type and page region. To examine a specific page region in the list of page regions configured for your page type, select it in the **Page regions** pane and click **Go To**. Styler will take you to that page region's definition.

Expert tip!

Every page on which you want to put content must have a main content region. If you're constructing an entire page with dynamic content (like a cover or back cover), those pages must have a main content region. Blank pages are the only page types for which a main content region is not required. You can leave blank pages empty, or you can add a static phrase like, "This page intentionally left blank." Fully generated pages with *dynamic content* are not considered blank.

Create page regions

Click the Page Regions tab to see the list of preconfigured page types from the OOTB module. You already know what's coming: the creation of a lot of page regions to match the page types that match the page sets we created.

You'll need a full set of regions to support all the possible page types. You created one of each of these page types for each of your page sets:

- Right pages
- Left pages
- First right pages

All of these page types have one or more of the following items:

- Header
- Footer
- Body

Use the same copy-paste-rename-move method to create the page regions you need.

Let's do it!

Click the **Page Regions** tab.

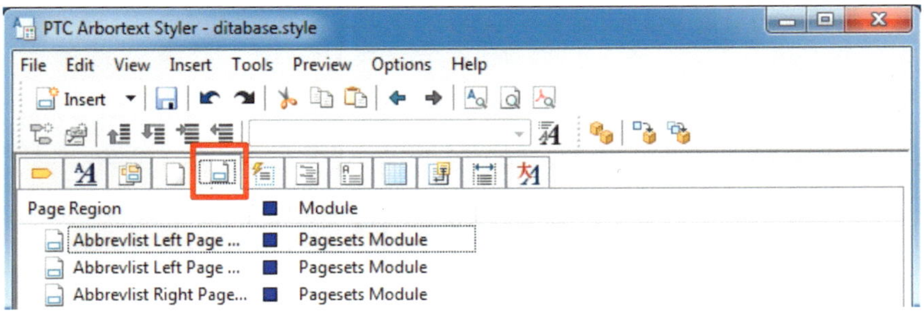

Copy these rules to your page set module:

- **Chapter Right Page Header**
- **Contents Right Page Header**
- **Index Right Page Header**

Copy and paste the **Abbrevlist *Left* Page *Footer*** to create these page regions:

- Chapter Left Page Footer
- Contents Left Page Footer
- Index Left Page Footer
- Appendix Left Page Footer
- Preface Left Page Footer

Copy and paste the **Abbrevlist *Left* Page *Header*** to create these page regions:

- Chapter Left Page Header
- Contents Left Page Header
- Index Left Page Header
- Appendix Left Page Header
- Preface Left Page Header

Copy and paste the **Abbrevlist *Right* Page *Footer*** to create these page regions:

- Chapter Right Page Footer
- Contents Right Page Footer
- Index Right Page Footer
- Appendix Right Page Footer
- Preface Right Page Footer

Copy and paste the **Abbrevlist *Right* Page *Header*** to create these page regions:

- Appendix Right Page Header
- Preface Right Page Header

Move all of the page regions you created to your page sets module.

When you're done, you should have this:

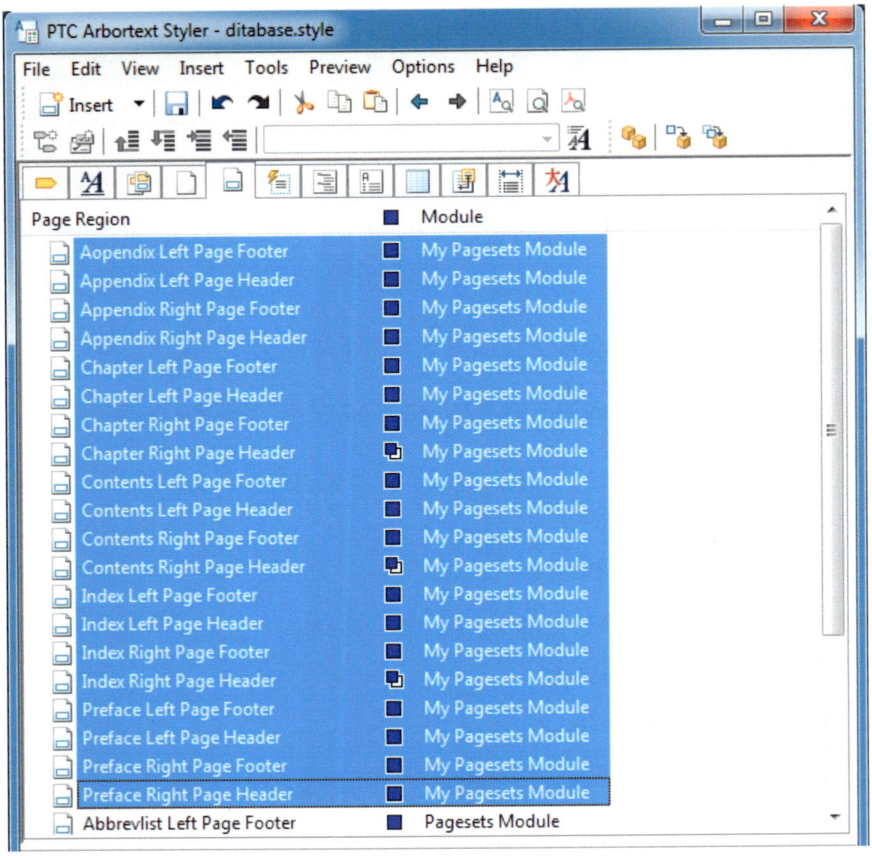

Save your stylesheet.

Examine a page region

Now that you have all the page regions created, take a closer look at one of them. Click **Chapter Right Page Header** in the upper panel, then click the **Text** category in the lower panel.

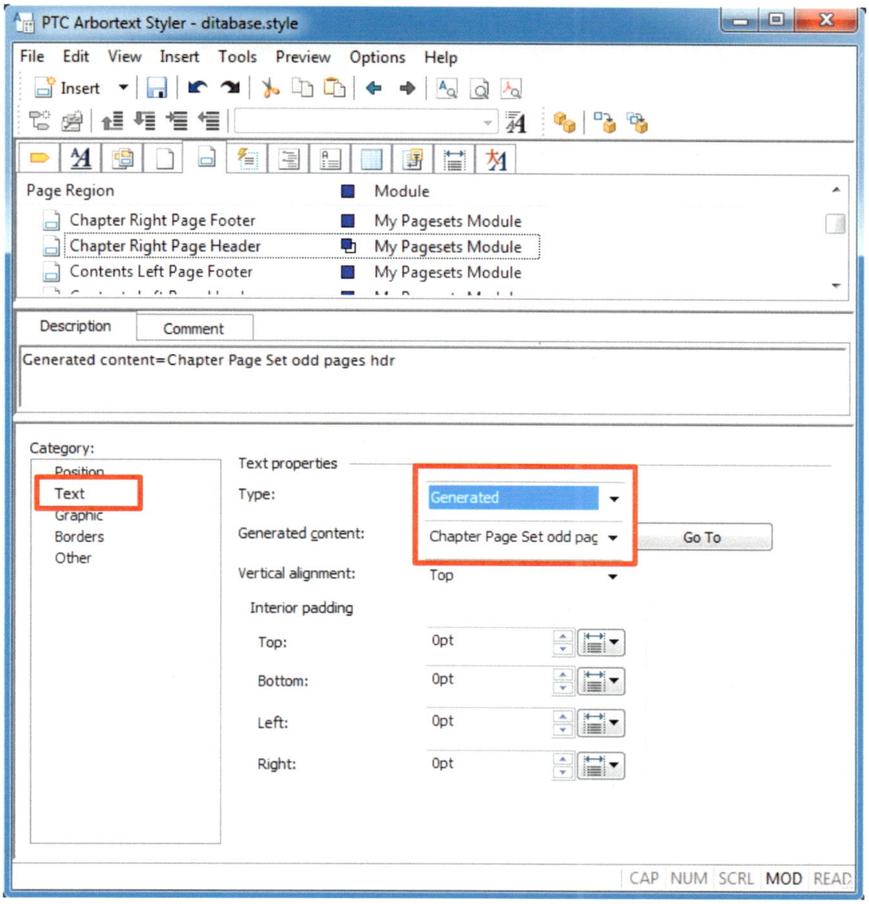

Page regions can have entirely generated content. If you want generated content, set **Text property→Type** to **Generated**. For this region, it's already set because we copied it from the OOTB stylesheet. Once you choose **Generated** for the type, you can select which **Generated content** definition you want to use from the dropdown menu.

Right now, the generated content we're using is **Chapter Page Set odd pages hdr**:

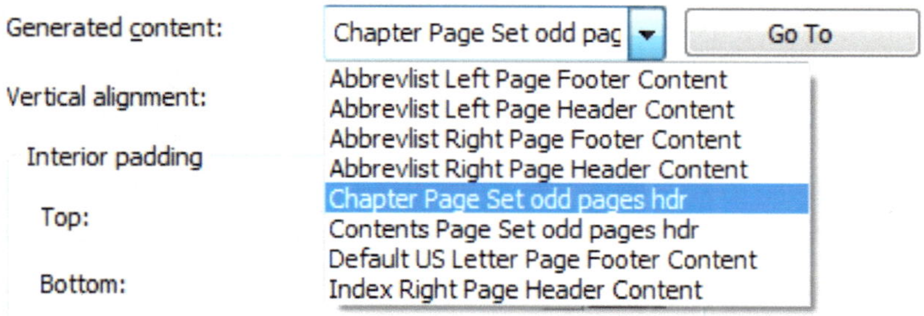

Take a look at a couple of the other regions you created. As you might expect by now, you have a variety of generated content items that aren't named consistently to match the page types, regions, or sets. Because of the way we copy and pasted, nearly all of them are identical.

You're going to change this too. You'll create new generated content and reconfigure the regions to match. Not all headers and footers are created equal. They frequently vary in different contexts. Chapter page headers might be different from index page headers, for example.

Don't worry. I'll show you what to do to create all the necessary generated content and how to configure them so they display the right content in the right places.

Page regions and generated content

A page region has a lot of capability. Headers and footers, for instance, are either static or generated, depending on their purpose within their page set and the greater document context.

It is common to find a one-to-one relationship between a page region and generated content. Between these two stylesheet components, you bring flexibility to your stylesheet. You can apply advanced formatting techniques when you bring it all together.

Create generated content

Generated content is dynamically constructed content that you can define and control. You can pull content from anywhere in your document into generated content. You have access to an XML editor when you are creating and formatting generated content, so you don't have to hand-code it.

You're going to create the template for all the headers and footers for all the page regions that support all the page types for the page sets you've created so far. While I've limited this step to the headers and footers, generated text supports much more than that. Once you're done, you'll have a better idea about what you can do with generated content. Headers and footers make a good introduction to the topic.

Let's do it!

Click the **Generated Content** tab.

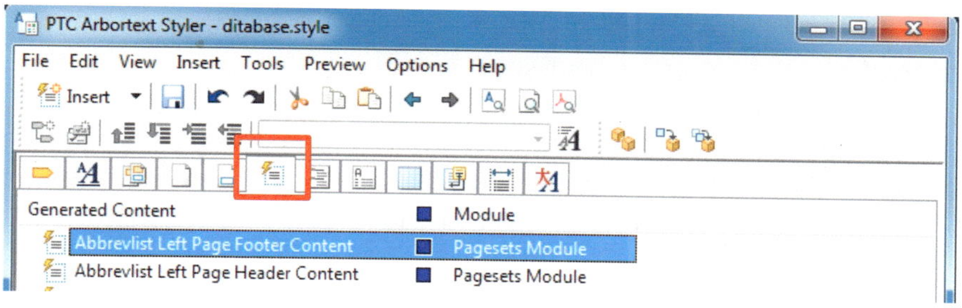

As with the other tabs, we have a small list here that doesn't have everything we want. Same song, different verse.

Copy these to your page sets module and rename them:

Copy	**Rename to**
Chapter Page Set odd pages hdr	Chapter Right Page Header Content
Contents Page Set odd pages hdr	Contents Right Page Header Content

Copy this one to your page sets module.

- Index Right Page Header Content

Copy and paste the **Abbrevlist Left Page Footer Content** and rename it to create these generated content rules:

- Chapter Left Page Footer Content
- Contents Left Page Footer Content
- Index Left Page Footer Content
- Appendix Left Page Footer Content
- Preface Left Page Footer Content

Copy and paste the **Abbrevlist Left Page Header Content** and rename it to create these generated content rules:

- Chapter Left Page Header Content

- Contents Left Page Header Content
- Index Left Page Header Content
- Appendix Left Page Header Content
- Preface Left Page Header Content

Copy and paste the **Abbrevlist Right Page Footer Content** and rename it to create these generated content rules:

- Chapter Right Page Footer Content
- Contents Right Page Footer Content
- Index Right Page Footer Content
- Appendix Right Page Footer Content
- Preface Right Page Footer Content

Copy and paste the **Abbrevlist Right Page Header Content** to create these generated content rules:

- Index Right Page Header Content
- Appendix Right Page Header Content
- Preface Right Page Header Content

Move all of the generated content you created to your page sets module.

When you're done, you should have this:

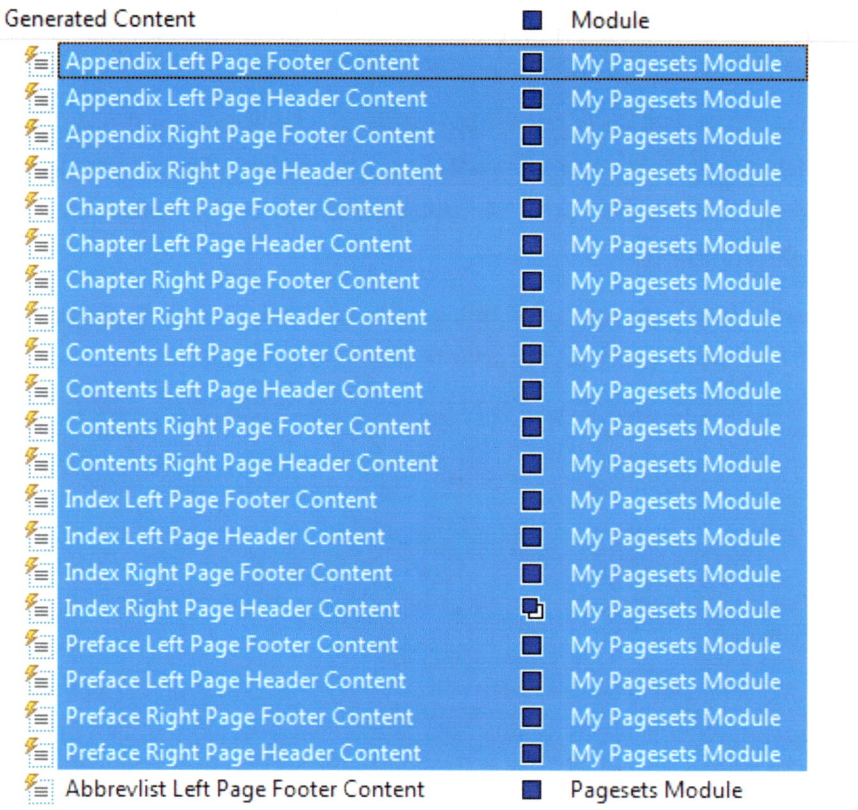

Save your stylesheet.

One last thing and we'll start working backwards.

Create a cross-reference format

When you're working with headers and footers, you need to understand the complex relationship between generated content and cross-references. Cross-reference formats are technically generated content because they are created automatically by the stylesheet. However, they are not configured under the **Generated Content** tab. Instead, they are configured under the **Cross Reference** tab.

While headers and footers can be configured lots of different ways, it's common to see a couple of things in the headers of a book. For example, a left page will often display the book title. A page might show a running header or, in the case of an appendix, only the label ("Appendix") and the appendix number ("A").

When you want to display a generated value in a header, you are actually retrieving the cross-reference content. The header text doesn't create a live link but it does display the same text you would see as if it were a cross reference in running text.

To get a better idea about how all this works, click the **Cross Reference** tab and take a look at what comes preconfigured.

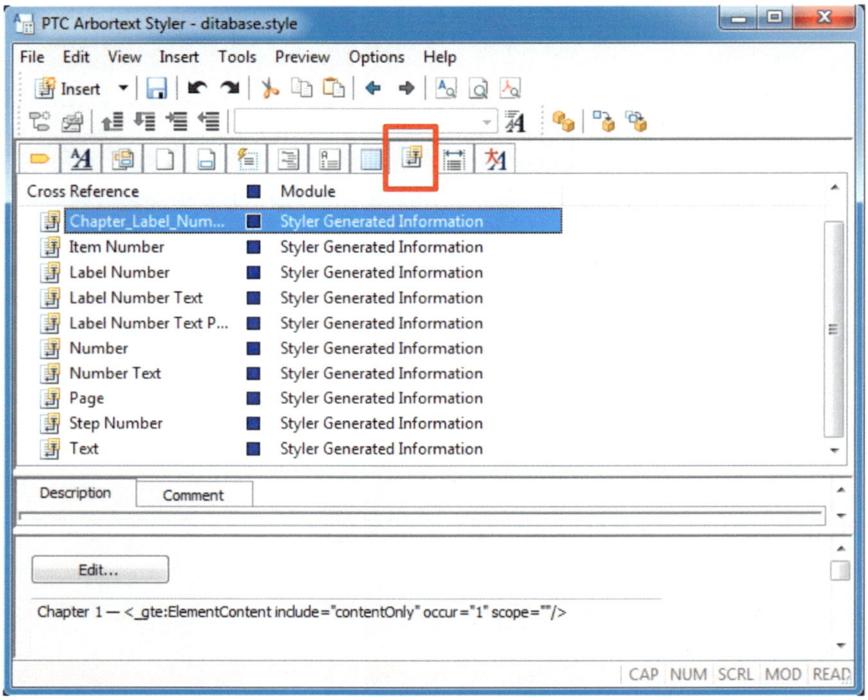

If you click each cross-reference rule in the upper panel, you can see in the bottom pane how each is configured.

Most cross-reference formats are covered already. **Appendix** is not. None of the existing formats will give us the generated word "Appendix" and the label number ("A"), without the title text.

For example, suppose this is your markup:

```
<appendix>
  <title>Top Arbortext resources</title>
```

Then each cross-reference format would display the following:

Cross-reference format	**What is displayed**
Chapter_Label_Number_Text	Chapter A — Top Arbortext Resources
Item Number	Item A
Label Number	A
Label Number Text	A Top Arbortext Resources
Label Number Text Page	A Top Arbortext Resources, page <#>
Number	A
Number Text	A Top Arbortext Resources
Page	, page <#>
Step Number	Step A
Text	Top Arbortext Resources

None of these settings matches the requirement. For example, **Chapter_Label_Number_Text** has a literal value. If you look closely at the lower panel, you see the word **Chapter**. Even if we apply this cross-reference format to an appendix, it will display the word **Chapter** instead of **Appendix**.

To get the right behavior for **Appendix**, you can duplicate the **Chapter_Label_Number_Text** format, copy it to your page set module, and change it. This is the format you will use to display the generated text in the header for the **Appendix Right Page Header Content**.

Let's do it!

Click the **Cross References** tab.

Copy and paste **Chapter_Label_Number_Text**.

Rename it to **Appendix_Label_Number**.

Move it to Module: **My StylerGenerated Module**.

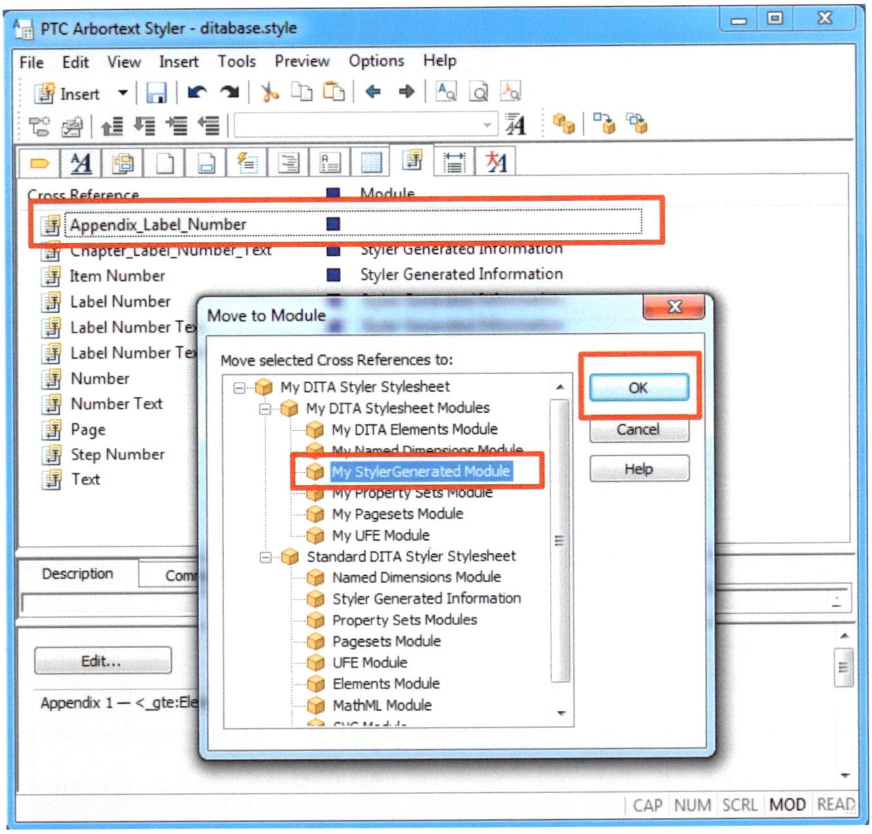

Select **Appendix_Label_Number** in the upper panel, then click the **Edit...** button in the lower panel. In the edit window, change **Chapter** to **Appendix**.

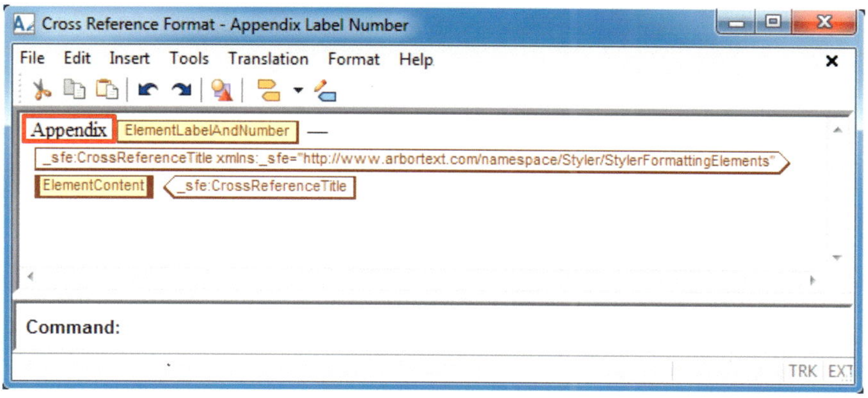

Close the window. Click **Yes** when you see the message asking if you want to apply your changes

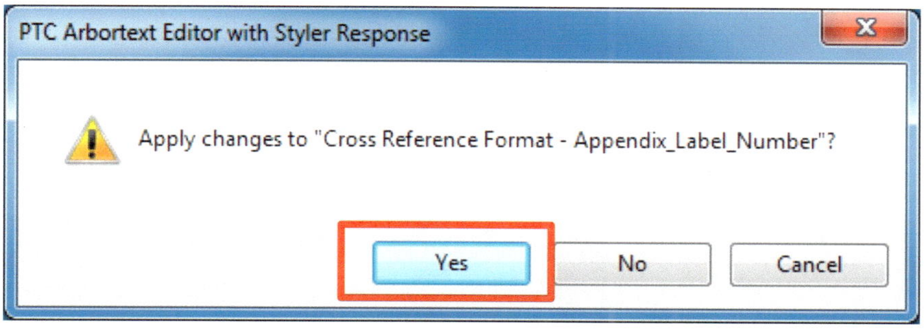

Save your stylesheet.

Now you have everything ready to start working backwards.

Configure the page regions

All the generated content has been created, so now you can configure the page regions.

The best way to proceed is to sort the list of page regions by module. That way, you can start at the top and work your way down.

Let's do it!

Click the **Page Regions** tab.

Click **Appendix Left Page Footer** in the upper panel.

Click **Text** in the **Category** pane in the lower panel.

Select **Appendix Left Page Footer Content** from the **Generated Content** text property dropdown list.

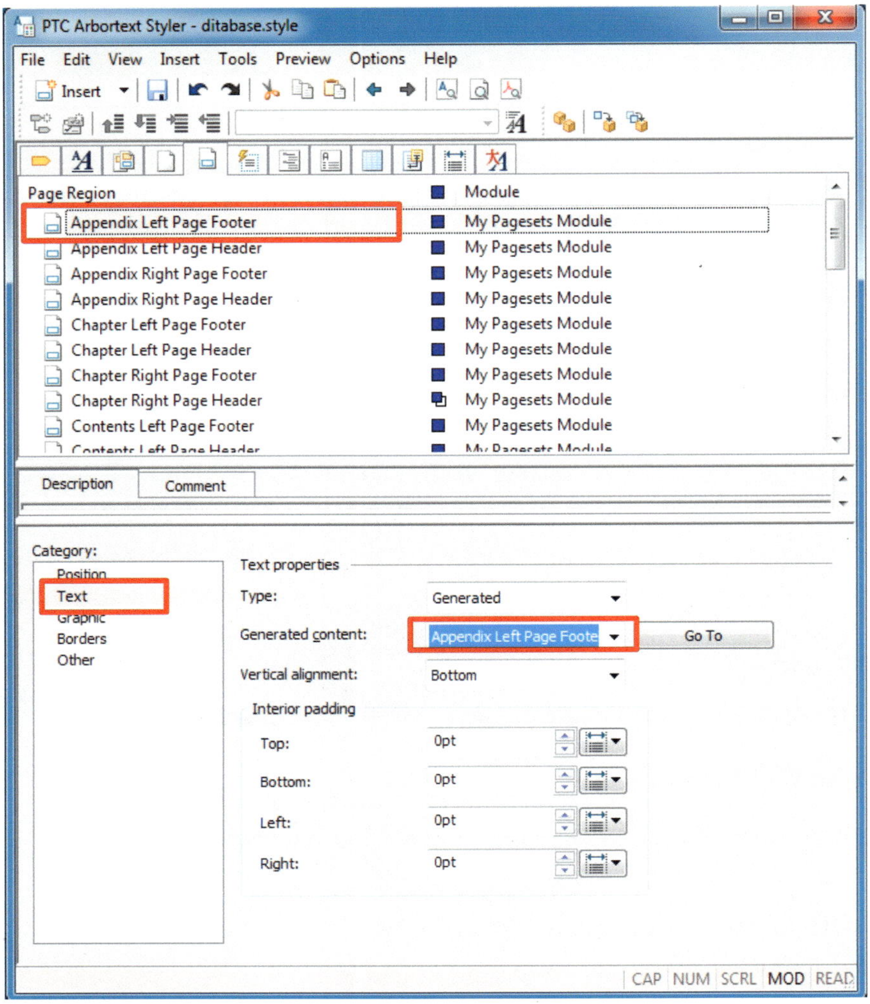

Do the same for all the page regions in the **My Pagesets Module**.

Match page regions with generated content like this:

Page Region	**Generated Content**
Chapter Right Page Header	Chapter Right Page Header Content
Chapter Right Page Footer	Chapter Right Page Footer Content

Page Region	Generated Content
Chapter Left Page Header	Chapter Left Page Header Content
Chapter Left Page Footer	Chapter Left Page Footer Content
Contents Right Page Footer	Contents Right Page Footer Content
Contents Right Page Header	Contents Right Page Header Content
Contents Left Page Header	Contents Left Page Header Content
Contents Left Page Footer	Contents Left Page Footer Content
Index Right Page Footer	Index Right Page Footer Content
Index Right Page Header	Index Right Page Header Content
Index Left Page Header	Index Left Page Header Content
Index Left Page Footer	Index Left Page Footer Content
Appendix Right Page Header	Appendix Right Page Header Content
Appendix Right Page Footer	Appendix Right Page Footer Content
Appendix Left Page Header	Appendix Left Page Header Content
Appendix Left Page Footer	Appendix Left Page Footer Content
Preface Right Page Header	Preface Right Page Header Content
Preface Right Page Footer	Preface Right Page Footer Content

Page Region	Generated Content
Preface Left Page Header	Preface Left Page Header Content
Preface Left Page Footer	Preface Left Page Footer Content

Save your stylesheet.

Configure page types

Click the **Page Type** tab, then sort by module.

Reconfiguring page types is a little more complex than reconfiguring page regions. When you're working with page types, the order in which regions are laid on a page matters. Remember that page regions are laid like layers on a page front to back

Work your way through all the page types in your module. Select each page type in the upper panel, then in the lower panel, add the regions that belong to the page. Then make sure the regions are ordered correctly.

Let's do it!

Configure your page types so they include the correct page regions.

Configure the first right pages

Click on **Appendix First Right Page** in the upper panel. In the lower panel, click **Add**.

Add **Appendix Right Page Footer**. Click the **up arrow** until it is at the top of the list.

Click **Abbrevlist Right Page Footer**, then click **Remove**.

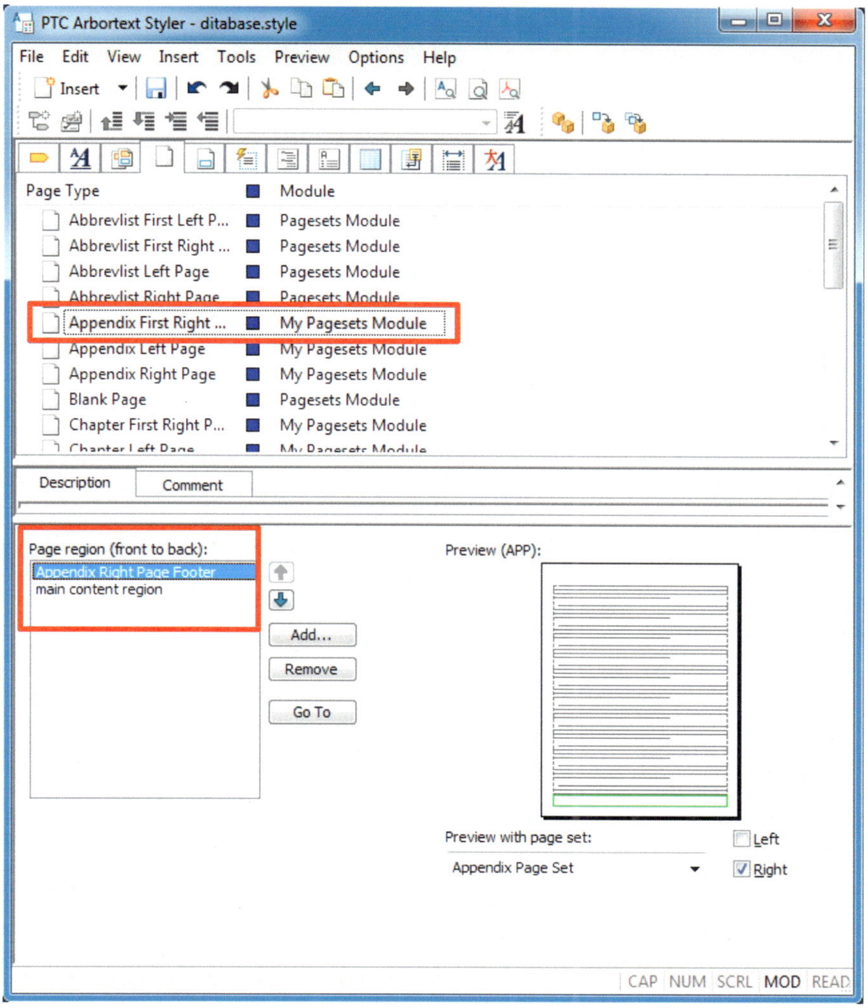

Repeat this process for all the First Right pages. First Right pages get only a footer. Each page type should be matched with the footer that corresponds to the page type, like this:

Page Type	Page Region(s)
Appendix First Right Page	Appendix Right Page Footer
Chapter First Right Page	Chapter Right Page Footer
Contents First Right Page	Contents Right Page Footer

Page Type	Page Region(s)
Index First Right Page	Index Right Page Footer
Preface First Right Page	Preface Right Page Footer

Configure the right pages

Right pages get both a header and footer. Be careful to match the header and footer region with the page type. Put the footer at the top (front) and the header at the bottom (back).

Page Type	Page Region(s)
Appendix Right Page	Appendix Right Page Footer Appendix Right Page Header
Chapter Right Page	Chapter Right Page Footer Chapter Right Page Header
Contents Right Page	Contents Right Page Footer Contents Right Page Header
Index Right Page	Index Right Page Footer Index Right Page Header
Preface Right Page	Preface Right Page Footer Preface Right Page Header

Configure the left pages

Left pages get both a header and footer. Be careful to match the header and footer region with the page type. Put the footer at the top (front) and the header at the bottom (back).

Page Type	Page Region(s)
Appendix Left Page	Appendix Left Page Footer Appendix Left Page Header
Chapter Left Page	Chapter Left Page Footer Chapter Left Page Header
Contents Left Page	Contents Left Page Footer Contents Left Page Header
Index Left Page	Index Left Page Footer Index Left Page Header
Preface Left Page	Preface Left Page Footer Preface Right Page Footer

Save your stylesheet.

Configure page sets

Configure your page sets so they include the correct page types.

Click the **Page Sets** tab and sort by module.

Click **Appendix Page Set** in the upper panel and **Page types** category in the lower panel.

Change each page type to the match the page type you created.

Page Type	Choose
Right pages	Appendix Right Page
First page when right	Appendix First Right Page

Page Type	**Choose**
Left pages	Appendix Left Page
First page when left	(Same as left pages)
Blank right pages	Blank Page
Blank left pages	(Same as blank right pages)

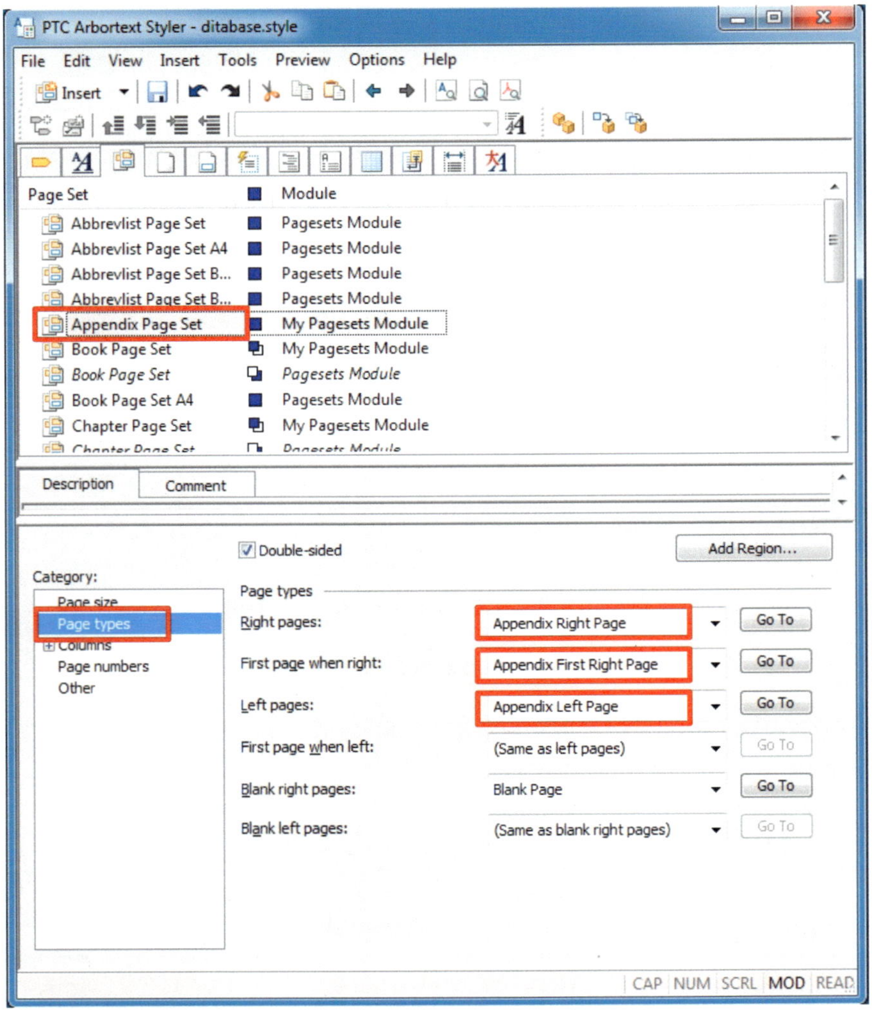

Do the same for the other page types listed below. Do not change the **Book Page Set**.

For the **Chapter Page Set**

Page Type	**Choose**
Right pages	Chapter Right Page
First page when right	Chapter First Right Page
Left pages	Chapter Left Page
First page when left	(Same as left pages)
Blank right pages	Blank Page
Blank left pages	(Same as blank right pages)

For the **Contents Page Set**

Page Type	**Choose**
Right pages	Contents Right Page
First page when right	Contents First Right Page
Left pages	Contents Left Page
First page when left	(Same as left pages)
Blank right pages	Blank Page
Blank left pages	(Same as blank right pages)

For the **Index Page Set**

Page Type	Choose
Right pages	Index Right Page
First page when right	Index First Right Page
Left pages	Index Left Page
First page when left	(Same as left pages)
Blank right pages	Blank Page
Blank left pages	(Same as blank right pages)

For **Preface Page Set**

Page Type	Choose
Right pages	Preface Right Page
First page when right	Preface First Right Page
Left pages	Preface Left Page
First page when left	(Same as left pages)
Blank right pages	Blank Page
Blank left pages	(Same as blank right pages)

Save your stylesheet.

Customize page layout (header and footer content)

Everything you need is now in place. All the components of the framework have been created and configured together. To get to this point, you simply duplicated and created placeholders and then configured the placeholders across the framework components.

Now that the framework components are in place, you can customize the individual components to fit your requirements. For example, you can change the headers and footers so they display what you really want.

Before we get started, it's important to understand page layout. I'll walk you through customizing one complete page set. Then I'll give you details about how to set up the others and identify points where the differences matter.

I'll start with chapter pages. This is how I want to configure them:

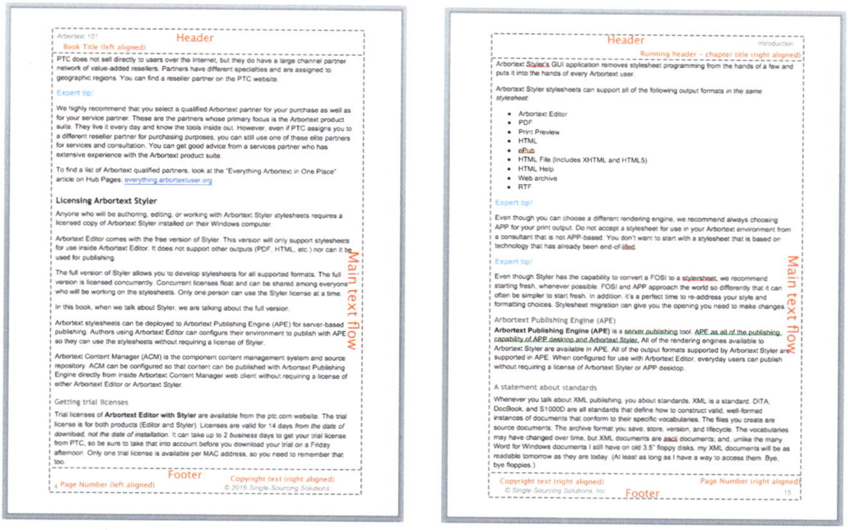

The diagram shows the regions—header, footer, body—and identifies what we want to show in each region. Before you start, you need to identify what elements in the source file contain the information you want to fetch and display in each region.

For the Left-Hand Page:

Footer	• Page number (left aligned, Arabic numeral) • Copyright symbol, copyright date, comma, company name (right aligned)
Header	• Book title (left aligned)

For the Right-Hand Page:

Footer	• Page number (right aligned, Arabic numeral) • Copyright symbol, copyright date, comma, company name (left aligned)
Header	• Running head (right aligned)

Expert tip!

You don't have to worry about setting up the **First Right Hand Page** because you took care of that when you configured the page regions. The first right hand page of a chapter has the same footer as other right pages and no header. As long as you configure the right hand page footer, the first right hand page will take care of itself.

Set the page numbering style

Displaying and formatting page numbering are two different tasks. First, you want to set the style. The page set determines page number style. Second, you want to add the page number to both the right and left page footers.

Click the **Page Set** tab.

Click the **Chapter Page Set** in the upper panel, then **Page Numbers** in the lower panel

Set the **Number** style to **1,2,3,…**

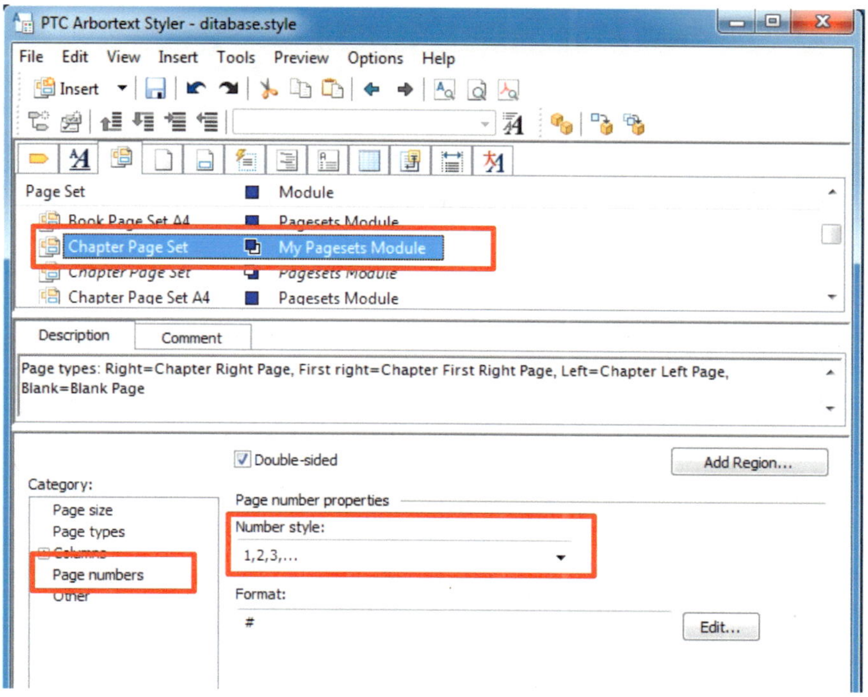

Set the page number style for both **Index Page Set** and the **Appendix Page Set** to **1,2,3,…** as well.

Select the **Contents Page Set**.

Change the page numbering style to **i,ii,iii,…**

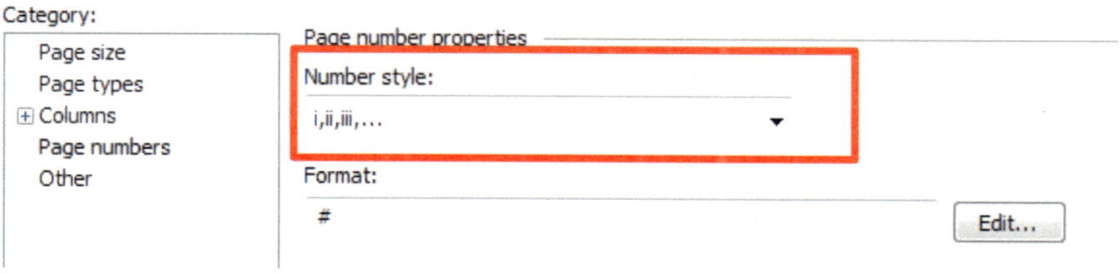

Set the page number style for **Preface Page Set** to **"i,ii,iii,…"** as well.

Save your stylesheet.

Now that the page number style is configured for your page sets, you need to make sure the page number is displayed in all the footers.

Left page footer

Click the **Generated Content** tab.

Select the **Chapter Left Page Footer Content** in the upper panel. In the lower panel, click **Edit**. Delete everything except the table markup. Place the cursor in the table cell.

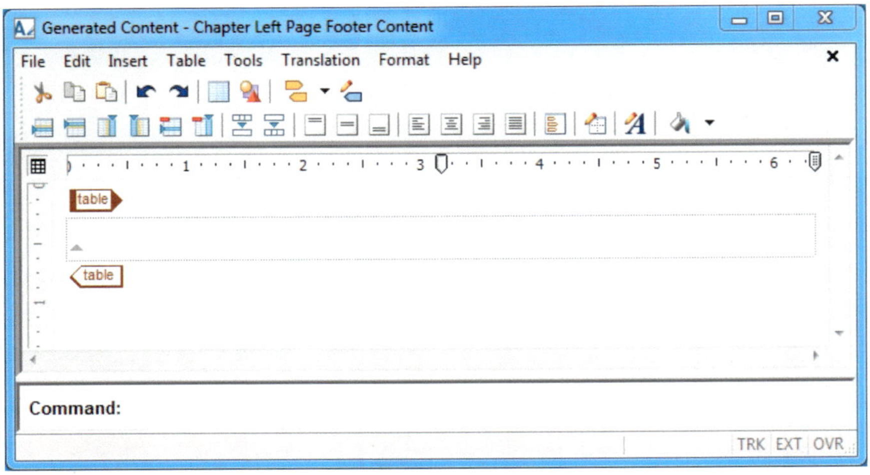

From the **Insert** menu, choose **Page Number**.

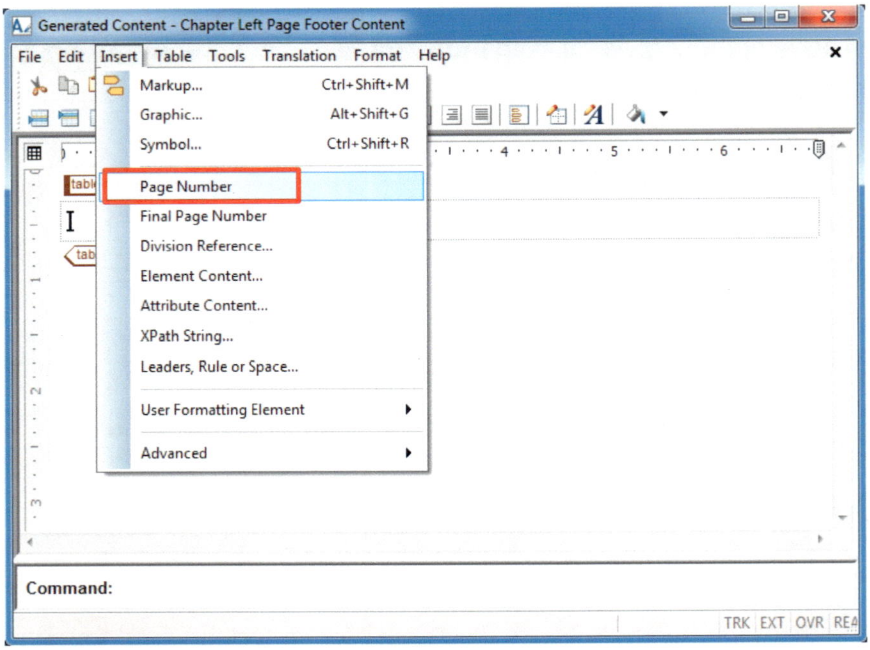

With the cursor still in the table cell, click the button to **insert column to right**.

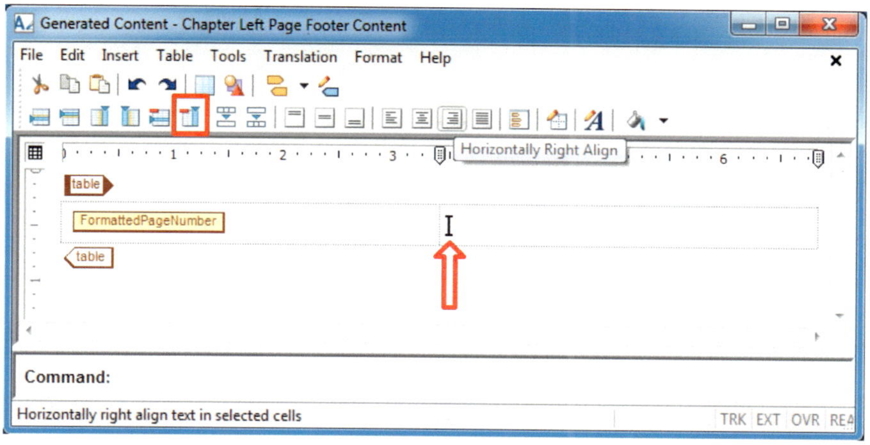

Place the cursor in the new cell. Click the **right align** button.

From the **Insert** menu, choose **Symbol...**

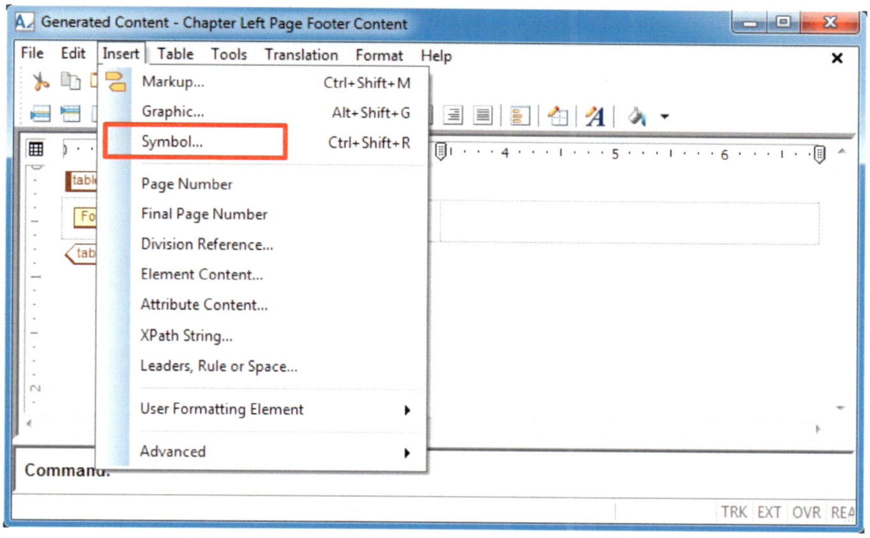

In the **Insert** Symbol window, choose the copyright symbol and then click **Insert**.

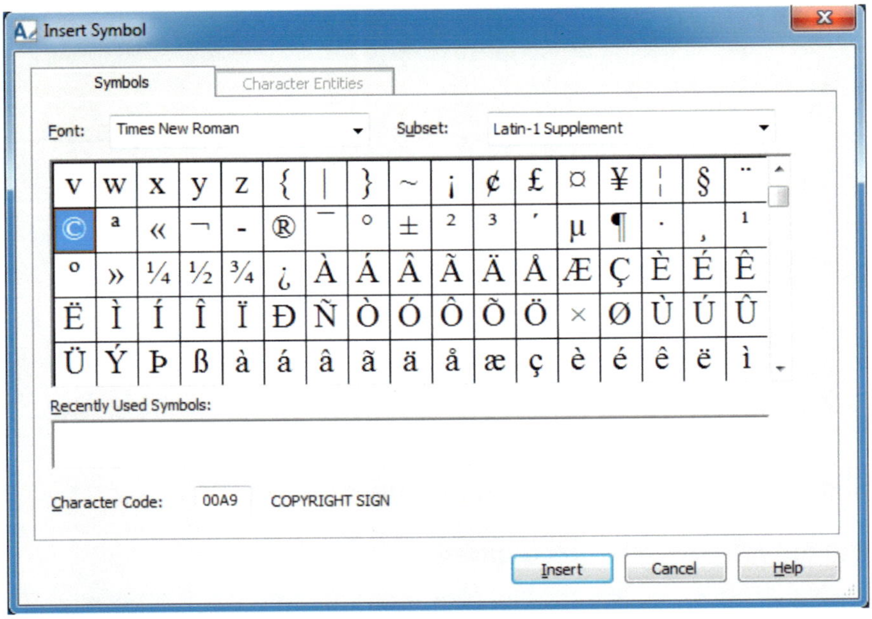

Close the symbol window. Add a space character, then type: **2016, My Company**.

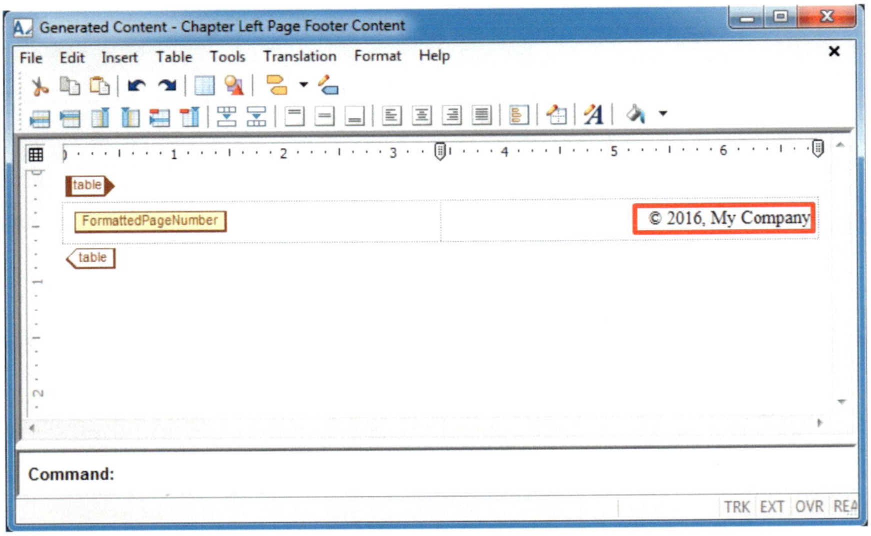

From the **File** menu, choose **Apply and Close**.

Save your stylesheet.

There's an easy way to replicate this footer content for the other Left Page Footer Content rules.

Click **Edit** for **Chapter Left Page Footer Content** to open the edit window again.

Press **Ctrl-A** to select all of the content in the window then copy it by pressing **Ctrl-C.**

Close this window. Paste this content into all the other Left Page Footers:

- Open **Appendix Left Page Footer Content**. Press **Ctrl-V** to paste. From the **File** menu, choose **Apply and Close**.
- Open **Contents Left Page Footer Content**. Press **Ctrl-V** to paste. From the **File** menu, choose **Apply and Close**.
- Open **Index Left Page Footer Content**. Press **Ctrl-V** to paste. From the **File** menu, choose **Apply and Close**.
- Open **Preface Left Page Footer Content**. Press **Ctrl-V** to paste. From the **File** menu, choose **Apply and Close**.

Save your stylesheet.

Expert tip!

There are many ways to construct content in the header and footers. In fact, using tables is not necessarily the optimum way. However, it is an easy way to get started and understand the tasks you are performing. It's also a good starting place for new stylesheet developers.

Expert tip!

Because the **Appendix Left Page Footer**, **Chapter Left Page Footer**, **Contents Left Page Footer**, and **Preface Left Page Footer** are all the same, you could have set each page type to use the same Footer region. While that would reduce complexity of your stylesheet and promotes reuse of stylesheet elements, we don't recommend it for beginning stylesheet developers, who can easily forget that something named ***Chapter* Left Page Footer** page region is also used by the **Appendix** left page type.

You can always simplify your stylesheet later, but having uniquely named items that match (page set names match page type names, which match page region names which match page generated content names) can be helpful when you're learning.

Right page footer

Select the **Chapter Right Page Footer Content** in the upper panel. In the lower panel, click **Edit**.

Delete everything except the table markup.

Repeat what you did for the left page footer, with opposite alignments:

1. Place the cursor in the cell, then click **align right**.
2. Insert page number.
3. Insert column to left.
4. Place the cursor in the new cell, then click **align left**.
5. Insert the copyright symbol ©, then type **2016, My Company**.

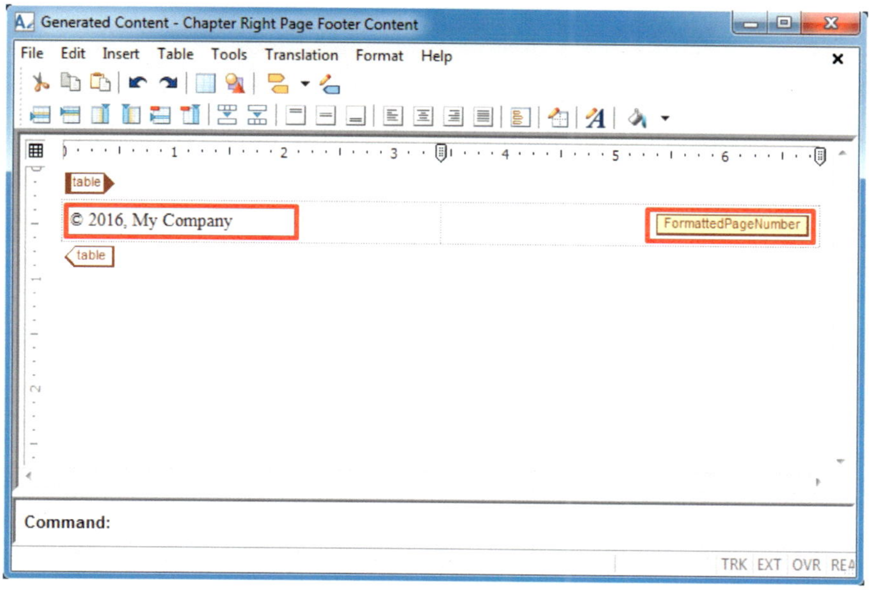

From the **File** menu, choose **Apply and Close**.

Save your stylesheet.

Select **Chapter Right Page Footer Content** then click the **Edit** button. Select all (**Ctrl-A**) and copy (**Ctrl-C**).

Close the edit window

Now paste into all the other Right Page Footer content items:

- Open **Appendix Right Page Footer Content**. Press **Ctrl-V** to paste. From the **File** menu, choose **Apply and Close**.
- Open **Contents Right Page Footer Content**. Press **Ctrl-V** to paste. From the **File** menu, choose **Apply and Close**.
- Open **Index Right Page Footer Content**. Press **Ctrl-V** to paste. From the **File** menu, choose **Apply and Close**.
- Open **Preface Right Page Footer Content**. Press **Ctrl-V** to paste. From the **File** menu, choose **Apply and Close**.

Save your stylesheet.

Expert tip!

A better way to handle the copyright date and company name is to fetch the values for these from the source document. As long as these values exist in your source document *in a position that you can predict accurately*, you can do exactly this.

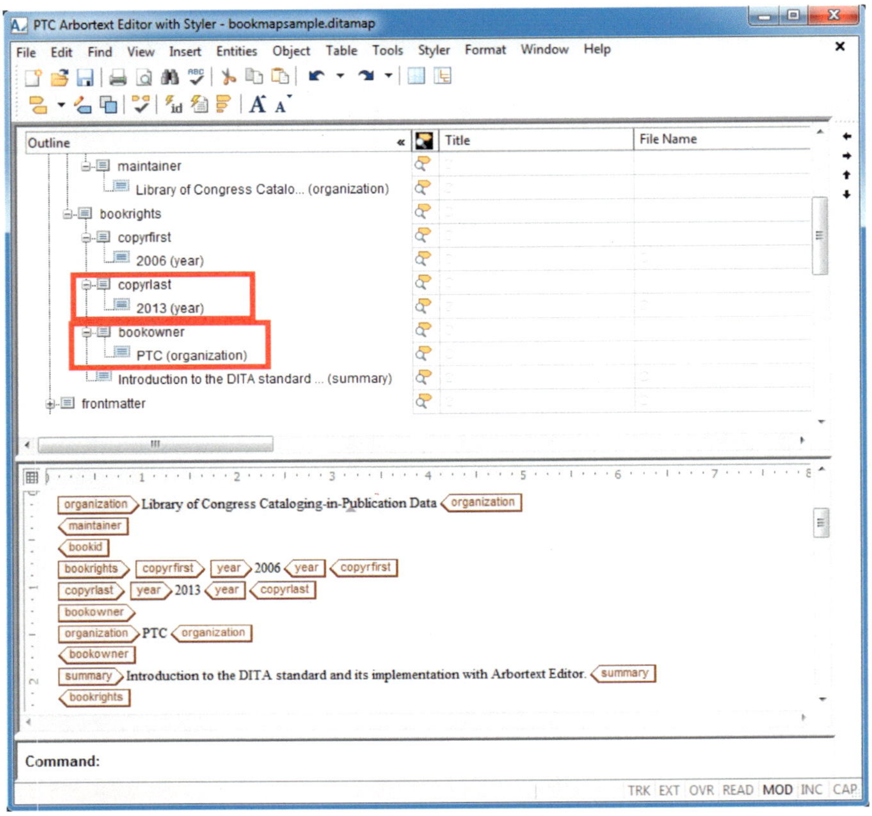

You can retrieve the *last* **year** element and the *first* **organization** element in the document, and you can display the element content in the footer generated content.

Instead of typing **2016, My Company**, you could do this:

After the copyright symbol, choose **Element Content** from the **Insert** menu. In the **Insert Element Content** window, set the following:

- For **Element selection,** choose **By name and occurrence**.
- For **Name**, choose **year** from the dropdown list.

- For **Specific occurrence within the document**, choose **last** from the dropdown list
- For Insert choose Only content.

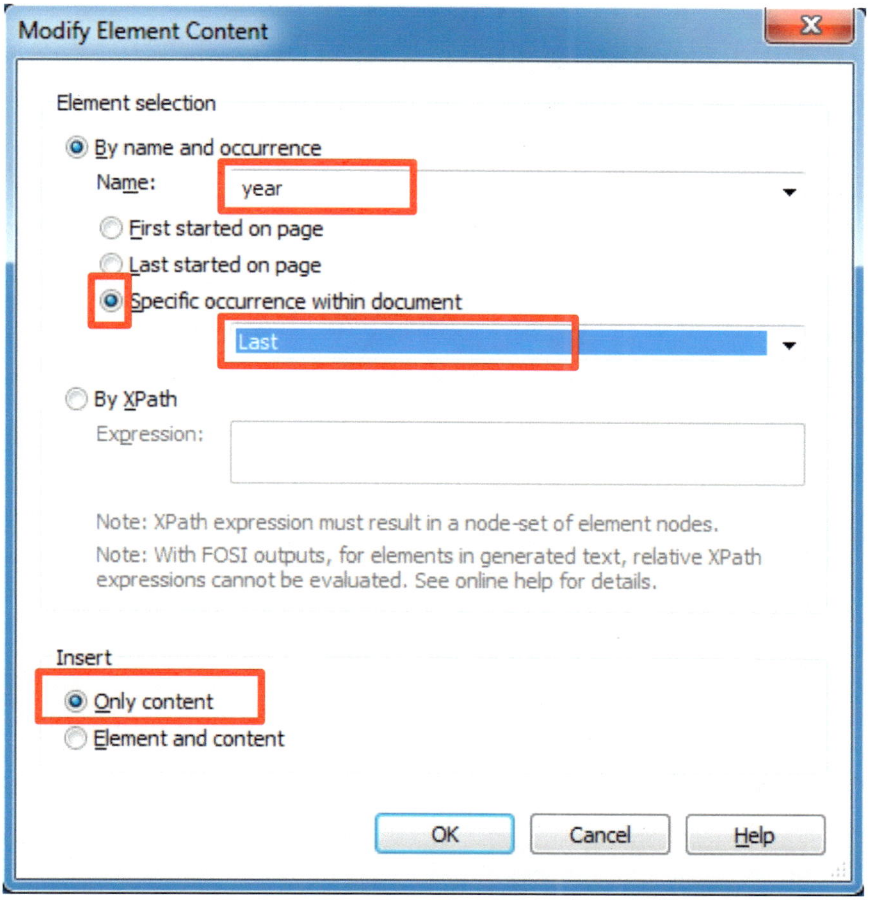

Repeat, choosing **organization** for **Name**. Keep all other settings the same:

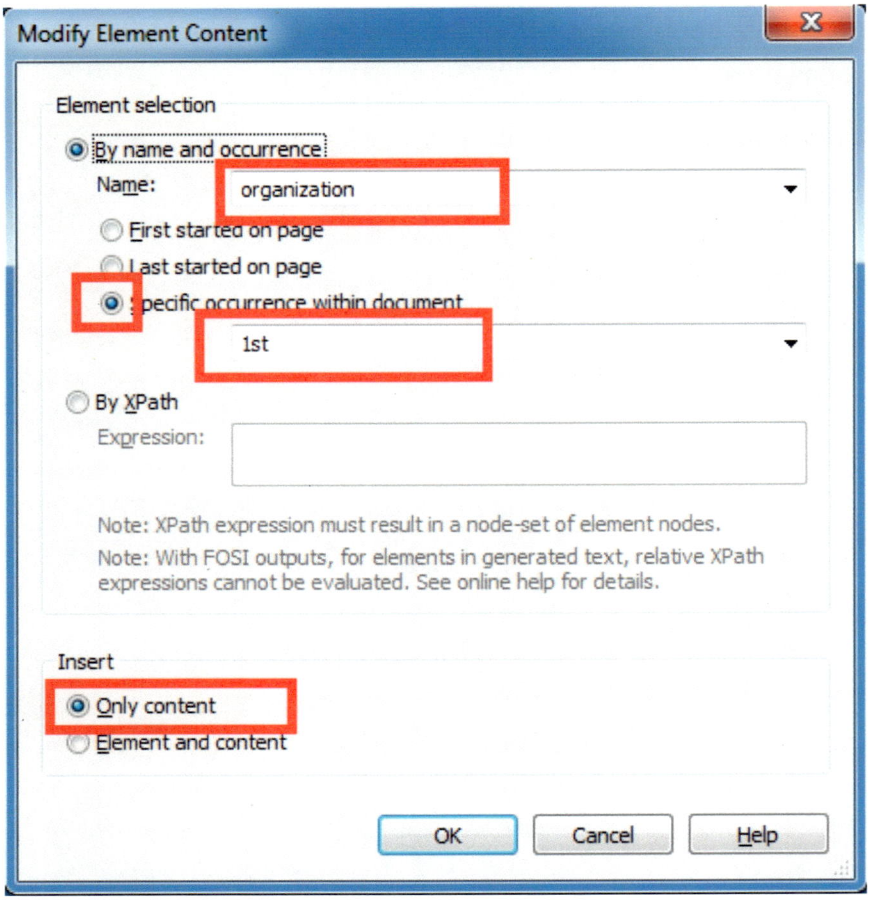

When you're done, this is what you'll have:

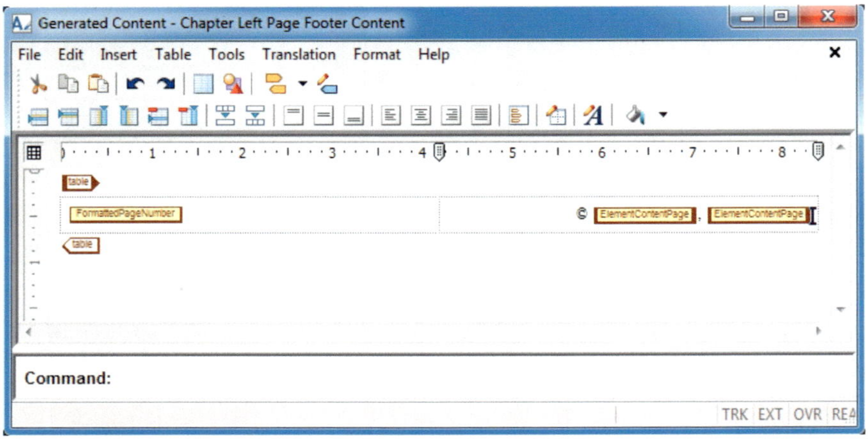

If you change the values in the **bookmap** the stylesheet will reflect the new values because it always displays the value specified in the **bookmap**. With this approach, your stylesheet will be more flexible, dynamic, and responsive.

You can use this method as long as you know the markup to expect—and that the markup is present in the source document. The stylesheet now has dependencies that you must communicate to your writers. Everything will work correctly only if the source has the content the stylesheet requires.

Left page header

The left page will display the book title in the header.

Click the **Generated Content** tab.

Select **Chapter Left Page Header Content** in the upper panel. In the lower panel, click **Edit**.

Delete everything except the table markup.

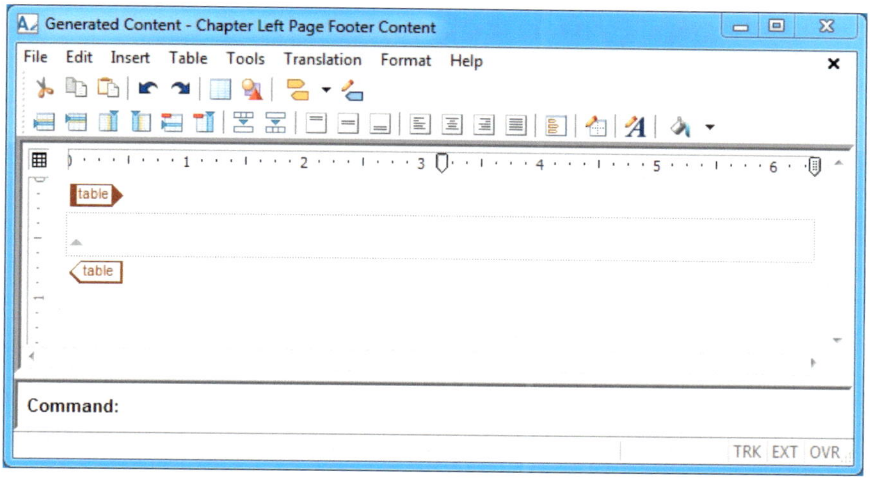

Place the cursor in the cell. Select **Insert→Element Content**.

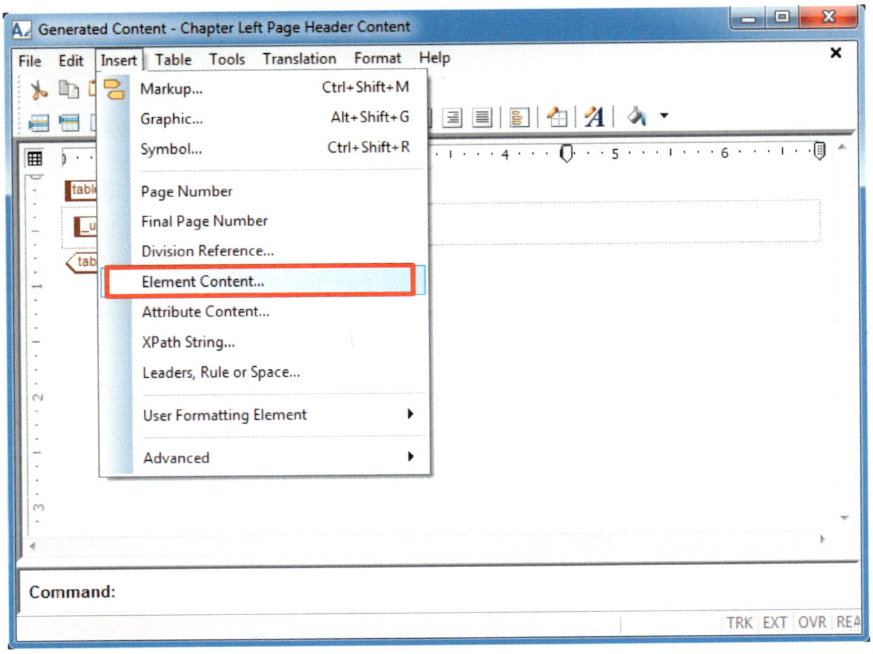

In the **Insert Element Content** window, choose the following settings:

- **Element selection**: By name and occurrence
- **Name**: mainbooktitle
- **Insert**: Only content

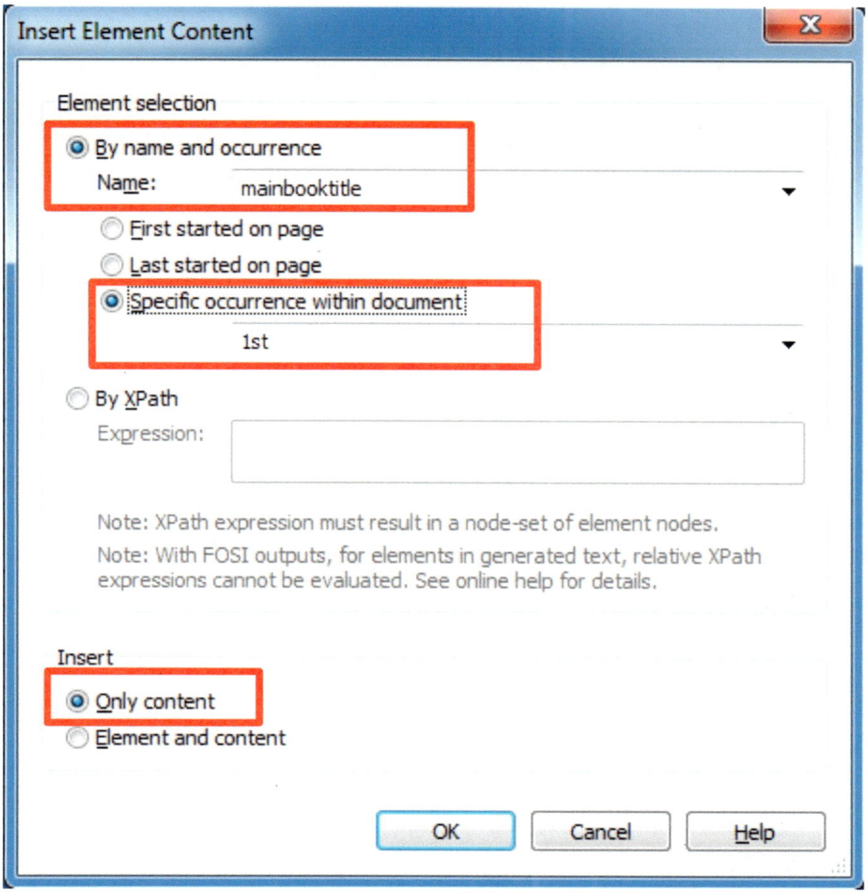

Like the footers, the header will display the title of the book as specified in the **mainbooktitle** element in the **bookmap**.

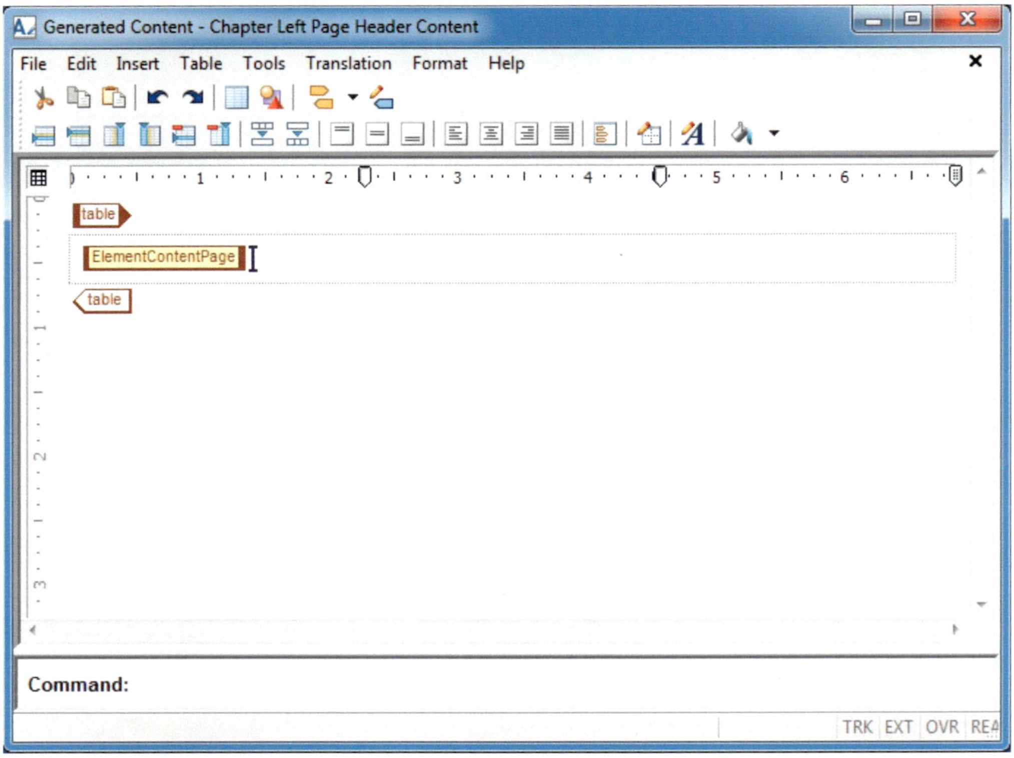

To finish, select **File→Apply and Close**.

Save your stylesheet.

Again, like you did with the footers, you can copy and paste the contents of the **Chapter Left Page Header Content** into the other **Left Page Header Content** items. Copy and paste this content into all of the following rules:

- Appendix Left Page Header Content
- Contents Left Page Header Content
- Index Left Page Header Content
- Preface Left Page Header Content

Select **File→Apply and Close**.

Save your stylesheet.

Right page header for chapter pages

Chapter right page header should display the running head. A running head changes as content is laid on the page. A running head shows the most recent division head appears on the page. It can be the first on the page or the last.

Select the **Chapter Right Page Header Content** in the upper panel. In the lower panel, click **Edit**.

Delete everything *except* the table markup.

Place the cursor in cell and click **align right**.

This time, choose **Insert→Division Reference...**.

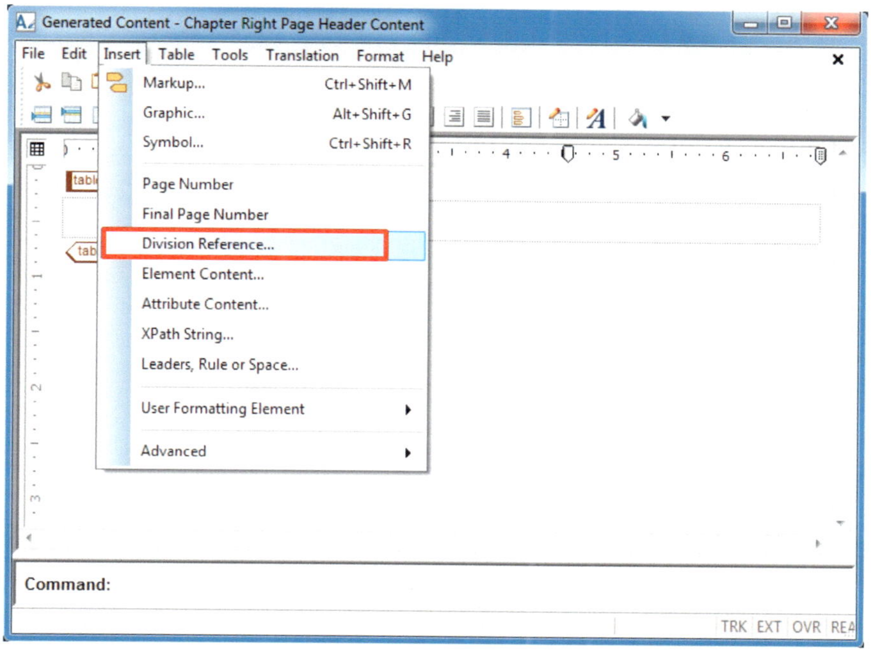

In the **Insert Division Reference** window, choose the following settings:

- **All divisions**: chapter
- **Selection criteria:** First division started on page
- **Cross Reference Format**: Text

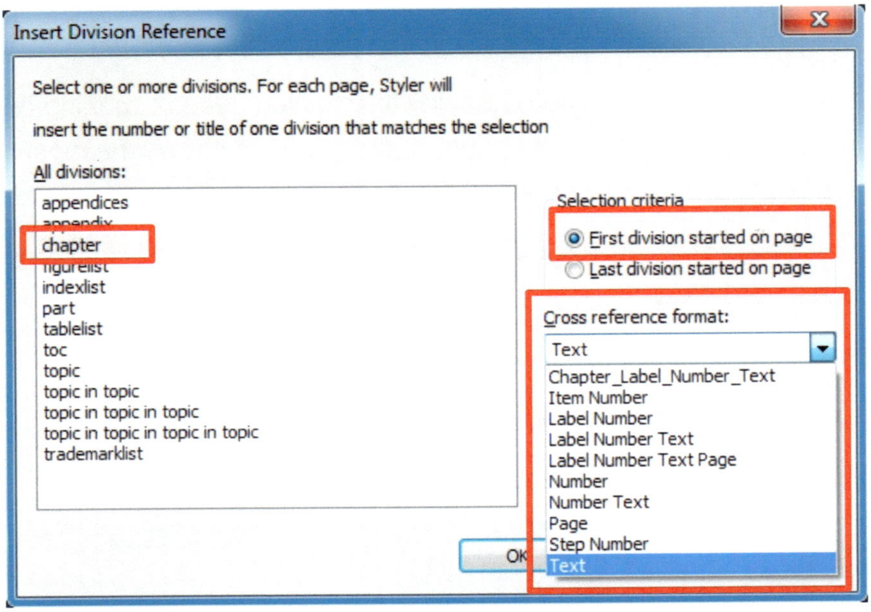

Click **OK** to close the window. You should see this:

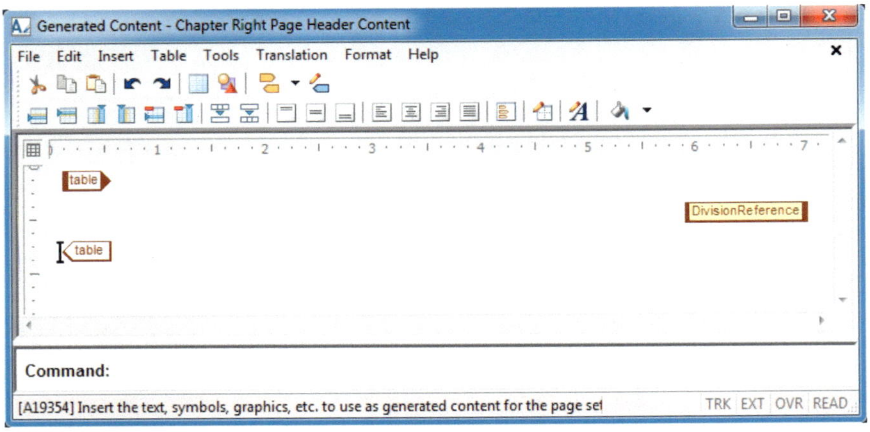

Select **File→Apply and Close**.

Save your stylesheet.

The right page header is different for all the page sets. It's the only one we can't copy and paste.

Expert tip!

Why **mainbooktitle**? Check document editor to know what element you want:

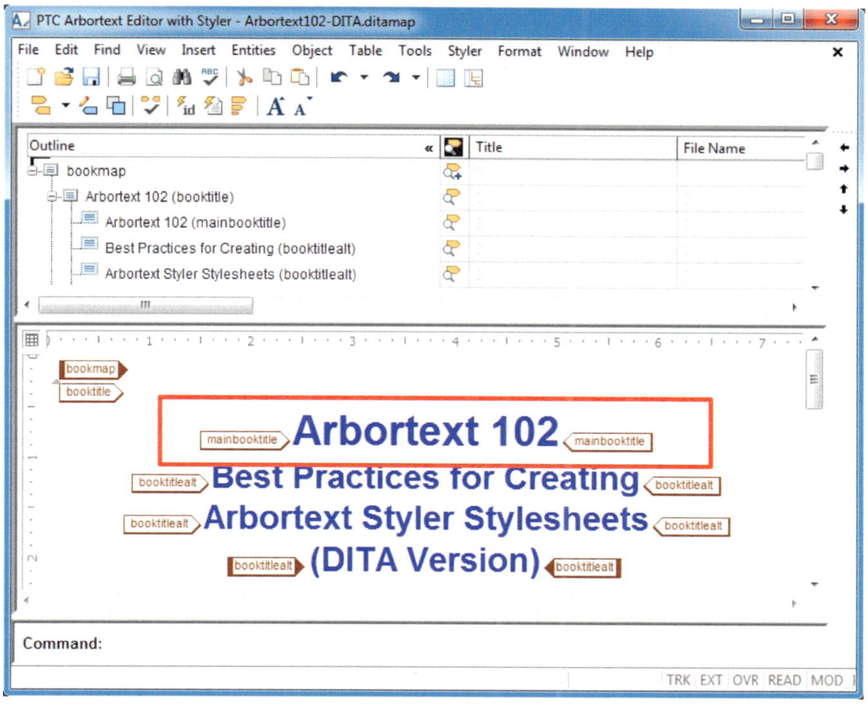

The book title you want to show in the left page header is the content in the **mainbooktitle** element.

Right page header for index pages

Index pages don't have running content because indexes are typically generated content. An index should be built automatically from all the index entries throughout the rest of the content.

Because an index doesn't have running head content and isn't always an appendix, it doesn't necessarily have a value we can retrieve and display. The header of an index page usually says simply **Index**.

Select the **Index Right Page Header Content** in the upper panel. In the lower panel, click the **Edit** button. Delete everything *except* the table markup.

Type **Index** and click **align right**.

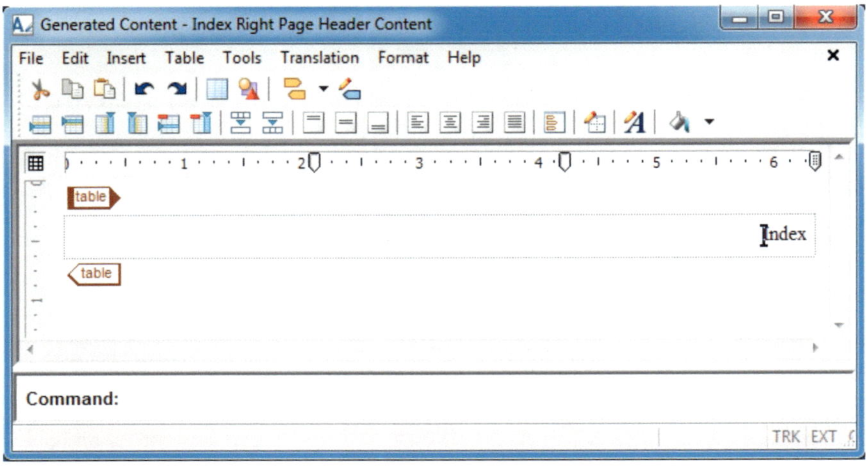

Select **File→Apply and Close**.

Save your stylesheet.

Right page header for contents pages

Like index pages, Contents pages tend to have a static title in the header.

Change the **Contents Right Page Header Content** to have static content like you did for the Index Right Page Header Content. Make the **Contents Right Page Header Content** say **Contents** instead of **Index**.

Right page header for preface pages

Preface pages have a running head like chapters, but they're not chapters. We will need to choose a different division reference.

Select the **Preface Right Page Header Content** in the upper panel. In the lower panel, click **Edit**. Delete everything *except* the table markup.

In the **Insert Division Reference** window, choose the following settings:

- **All divisions**: preface
- **Selection criteria:** First division started on page
- **Cross Reference Format**: Text

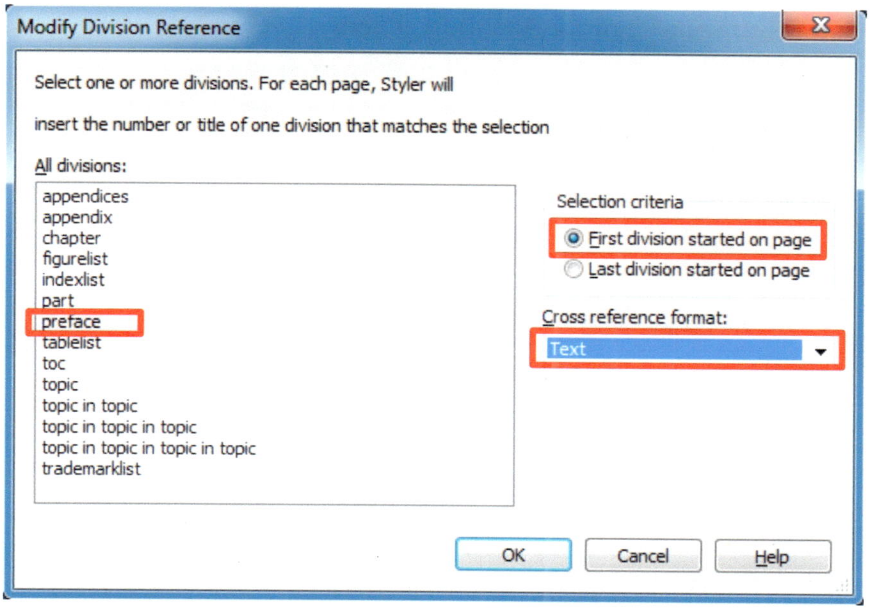

Select **File→Apply and Close**.

Save your stylesheet.

Right page header for appendix pages

- Appendix pages are often a combination of what you find on index pages and preface pages.
- Like index pages, they have a static header: Appendixes show the Label and Number ("Appendix A") in the header instead of the running head. You already set this up when you configured the appendix cross-reference format.
- Like preface pages, we can't use the same division reference as chapter (or preface). We have to use the appendix element as our division reference

Select **Appendix Right Page Header Content** in the upper panel. In the lower panel, click **Edit**. Delete everything *except* the table markup.

In the **Insert Division Reference** window, choose the following settings:

- **All divisions**: appendix
- **Selection criteria:** First division started on page

- **Cross Reference Format**: Appendix_Label_Number

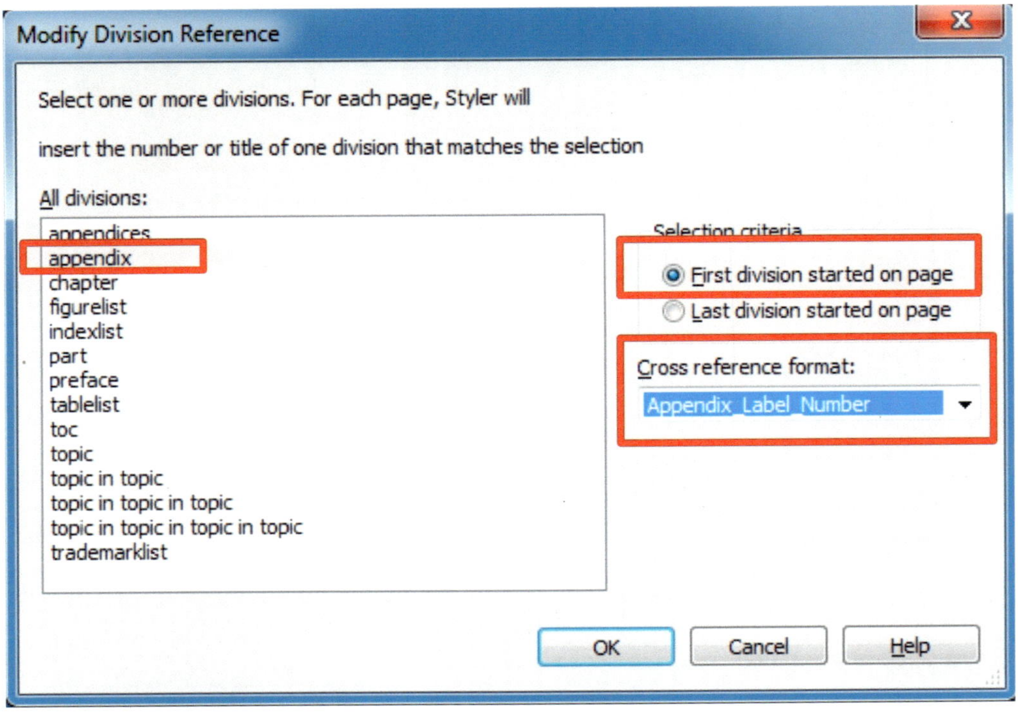

Select **File→Apply and Close**.

Save your stylesheet.

Now that your pages are set up, you need to record page set changes. You do that in the **Elements** tab.

Expert tip!

There are other parts of the book to handle, such as the cover—but we always recommend that you work on the cover last. It's common to want to do the cover first. It's right there in front of you. By working out the framework first, you will reduce the risk of publishing a book with structural problems. If you don't have the framework set up before working on the cover, you run the risk of introducing unexpected or conflicting changes when you add the rest of the framework later. You can lose yourself in a cover page, by endlessly tweaking it in response to intersecting framework changes.

We recommend completing the majority of the book publishing tasks before you start working on the cover page or any mostly generated page. It's a good idea to tackle the majority of formatting tasks first and do specialty work last.

Chapter 5. Working with element rules

Topics Covered in this Chapter

♦ Element rules and style types
♦ Element rules and output-specific formatting
♦ Element rules and page handling
♦ Configure page handling

An element rule encapsulates all the formatting properties for a doctype element supported by the stylesheet. Element rules are found under the **Elements** tab.

Most elements have one set of formatting properties. (A paragraph is a paragraph is a paragraph.) But some elements have different formats, depending on the context in which they appear. A title in a chapter (a first-level head) might look different from a title in a section in a chapter (a second-level head), and both might look different from a figure title. In the source document, all three of these are **title** elements but they all appear in different contexts:

- title in chapter
- title in section
- title in figure

You can add any number of contexts to an element rule to handle the different cases and formatting properties for elements with the same name (like **title**). You create a context for an element rule to define the unique situation. Then you apply the unique formatting properties to the context. When the rendering engine encounters a title in a specific context, it will use the properties attached to the context rule and not the general element rule. The rendering engine pays attention to the type and the context in which that element is used.

In addition to defining unique contexts, you can add conditions to a context that enable you to further refine the circumstances under which the context appears. You can apply conditional tests to an element rule and context.

For example, contexts allow you to set the formatting for a title in a section to look different from a title in a section that is for expert users only. You can test for the attribute value or define a complex XPath test. The results of the condition test determines which formatting properties should be applied to the specific element in the source document.

Contexts and conditions give you the flexibility to define the circumstances under which the stylesheet applies formatting properties to an element.

Take a look at the **toc** element rule as an example. Locate the **toc** element rule in the **Element** tab.

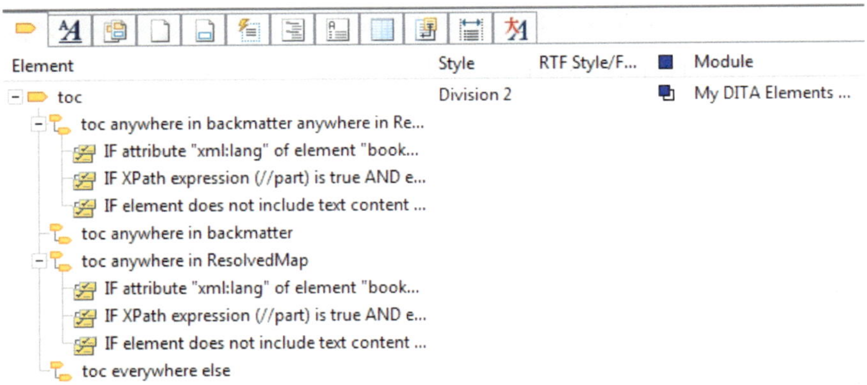

The **toc** element rule has several child rules. Each child rule that starts with the double yellow tag icon is a *context*. Each of the child rules that start with the double yellow block is a *condition*. The **toc** rule in the default stylesheet (shown) has four contexts. Two of those contexts have three conditions. Contexts and conditions can be complex or simple, depending on your needs.

Expert tip!

You can jump down the list of rules in the upper panel by pressing a key on your keyboard. The UI will jump to the first rule with that letter. To find **bookmap**, press the **b** key to get to the first element that begins with the letter b, then scroll down to find **bookmap**. You can use this technique anywhere in the upper panel.

Element rules and style types

Every element rule is assigned a basic style type.

Style types represent the categories of predefined styles available in Arbortext Styler. You can assign a style type to new element rules you add to your stylesheet or to user formatting elements (UFEs). An element rule must have a basic style type before formatting properties can be applied to the rule.

The list of predefined styles includes these:

- Block
- Cross reference
- Custom table
- Definition list
- Division
- Document
- Footnote
- Formal Block
- Graphic
- Hidden
- Index
- Index term
- Inline
- Link
- Link target
- List – Bulleted
- List – Numbered
- List item
- No style
- Paragraph
- Preformatted
- Table of Contents
- Title

By default, when you create a new element rule, the style is **Unstyled**. Unstyled elements cannot have styling applied to them. You must assign a predefined style before you can add formatting properties to an element rule.

By choosing a predefined style type for your element rule, you indicate to Arbortext Styler a profile for that element's usage in documents that are formatted with this stylesheet. For example, if you set the **preface** element's style as **Division**, then the preface element will be treated by the stylesheet as if it is equivalent to a **Chapter**, **Part**, or other major partition of your document structure.

You can assign a style type to an element rule by selecting the element rule, then going to the **Edit** menu. Choose **Style** and the style you want to apply to the element rule.

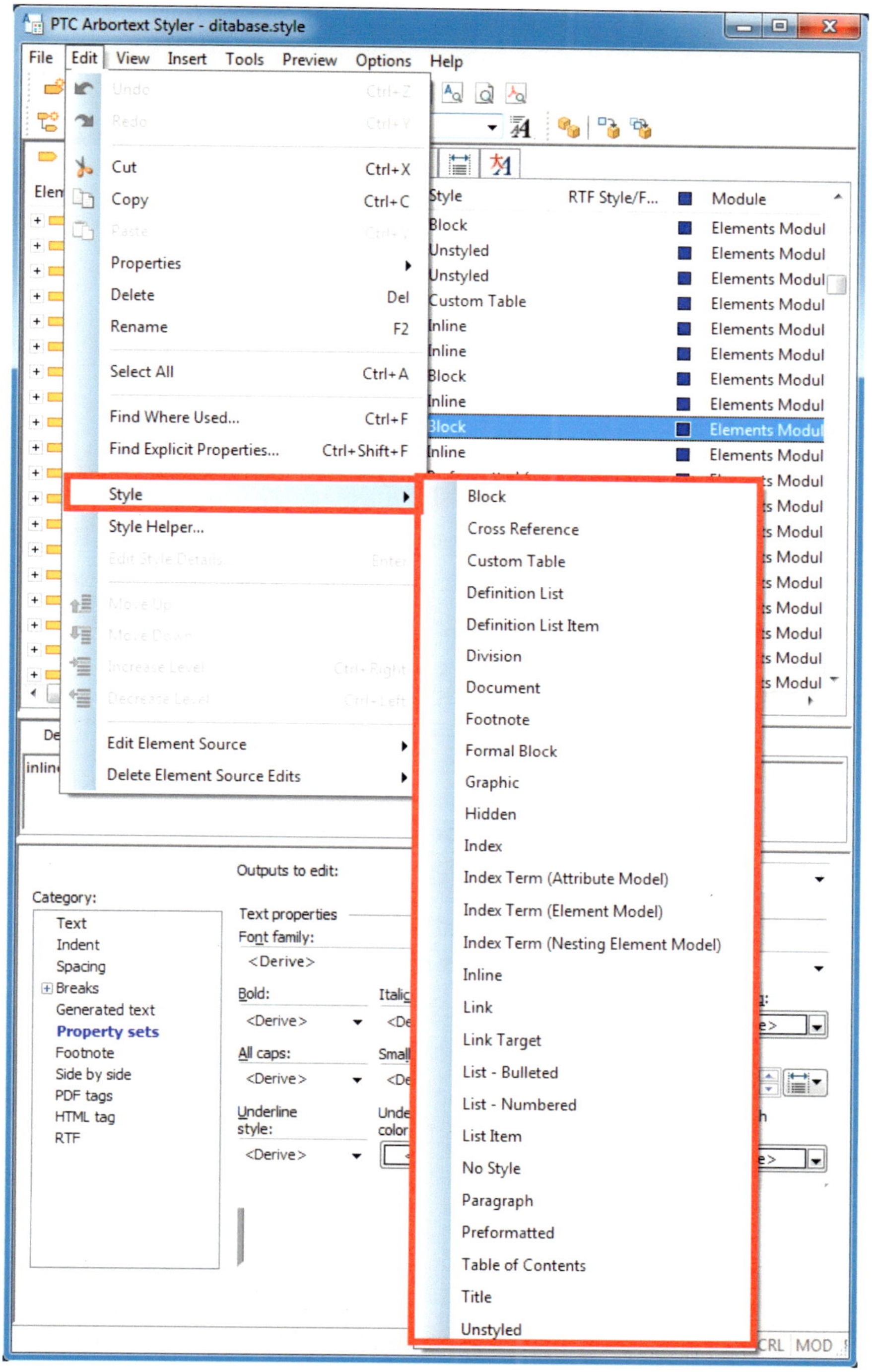
PTC Arbortext Styler - ditabase.style
File
Edit
View
Insert
Tools
Preview
Options
Help
Undo Ctrl+Z
Redo Ctrl+Y
Cut Ctrl+X
Copy Ctrl+C
Paste Ctrl+V
Properties
Delete Del
Rename F2
Select All Ctrl+A
Find Where Used... Ctrl+F
Find Explicit Properties... Ctrl+Shift+F
Style
Style Helper...
Edit Style Details Enter
Move Up
Move Down
Increase Level Ctrl+Right
Decrease Level Ctrl+Left
Edit Element Source
Delete Element Source Edits
Block
Cross Reference
Custom Table
Definition List
Definition List Item
Division
Document
Footnote
Formal Block
Graphic
Hidden
Index
Index Term (Attribute Model)
Index Term (Element Model)
Index Term (Nesting Element Model)
Inline
Link
Link Target
List - Bulleted
List - Numbered
List Item
No Style
Paragraph
Preformatted
Table of Contents
Title
Unstyled
Style
RTF Style/F...
Module
Elements Modul
Category:
Text
Indent
Spacing
Breaks
Generated text
Property sets
Footnote
Side by side
PDF tags
HTML tag
RTF
Outputs to edit:
Text properties
Font family:
<Derive>
Bold:
Italic
All caps:
Underline style:
MOD

Expert tip!

An element rule can have only one predefined style type assigned. All contexts and conditions have the same style type as the one assigned to their parent element rule.

Element rules and output-specific formatting

In addition to contexts and conditions, you can limit formatting properties to a single output format.

Before you can change page handling in an element rule, you must make sure your changes apply to the correct output format. Arbortext Styler supports multiple output formats at the same time in the same stylesheet. Whenever you work on an element rule, you must check the center panel to see if the rule has special settings for different output formats. If it does, be sure to select the output format you mean to affect before making any changes.

For example, look at the **bookmap** element. It is configured to handle all the different output types.

Click the **Element** tab and locate the **bookmap** element rule. If you look at the center panel of the stylesheet editor window, you'll see all the output formats listed in the **Description** tab. If you see an output format listed in the description, you know that this rule has settings tied to that output format in this stylesheet. If you want to effect that output format, you must change the selection in the **Outputs to edit** dropdown list before making your changes. Changes in the lower panel always apply to whichever output format is active in the dropdown list.

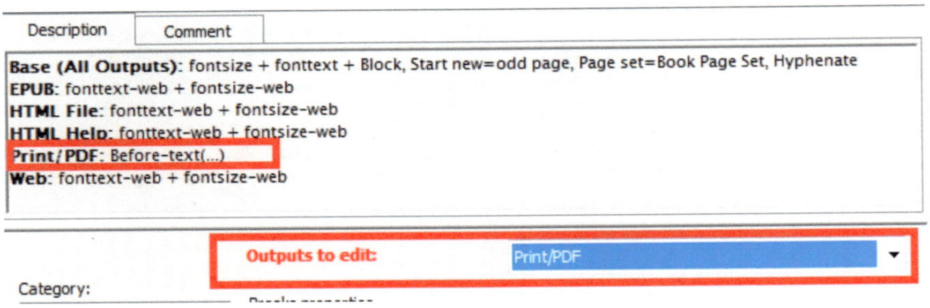

If you want to change the page set used for Print/PDF layout, you have to change the **Outputs to edit** dropdown list to **Print/PDF** before continuing.

If the **Description** tab is empty, you can either specify the output format you want to apply changes to, or you can leave it set to **Base (All Outputs)**, and the changes will apply to all output formats and the document editor window.

In the lower panel, the Arbortext UI gives you visual cues to identify which properties are set on any specific element rule. When a property has a manual setting change, the property name and the name of the category to which it belongs appear blue.

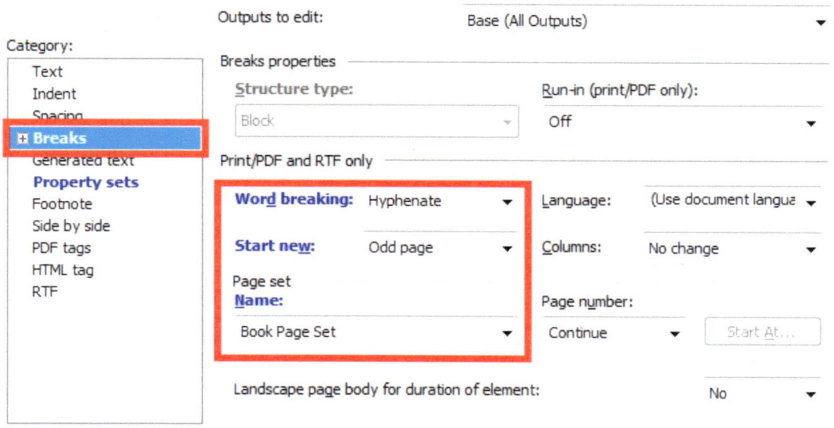

Blue means the property has been changed manually in this rule.

When you're troubleshooting your stylesheet, looking closely at the categories and properties in blue, along with the output format settings, can often help you identify the problem.

Expert tip!

When the description box is empty, the rule has no settings that are specific to an output type. The outputs dropdown menu should show **Base (All Outputs).** Any changes you make to the rule will apply to all outputs.

Element rules and page handling

If you have all your page sets configured, you still need to make sure that the right page set is executed at the right time. A page set is triggered by an element rule. A trigger indicates to the stylesheet to start using a specific page set.

You tell the stylesheet when to start a new page set, which page set to start, and how it should start by associating a page trigger in an element rule. You control page set execution by indicating the page set change in the element rule. The element rule riggers the page set change.

Whenever an element is encountered during processing, the stylesheet can start or stop a page set by starting a new right-hand or left-hand page. The first step in defining page set execution is to identify the relationship between elements and page sets. Start by looking at the sample source file (the XML input document):

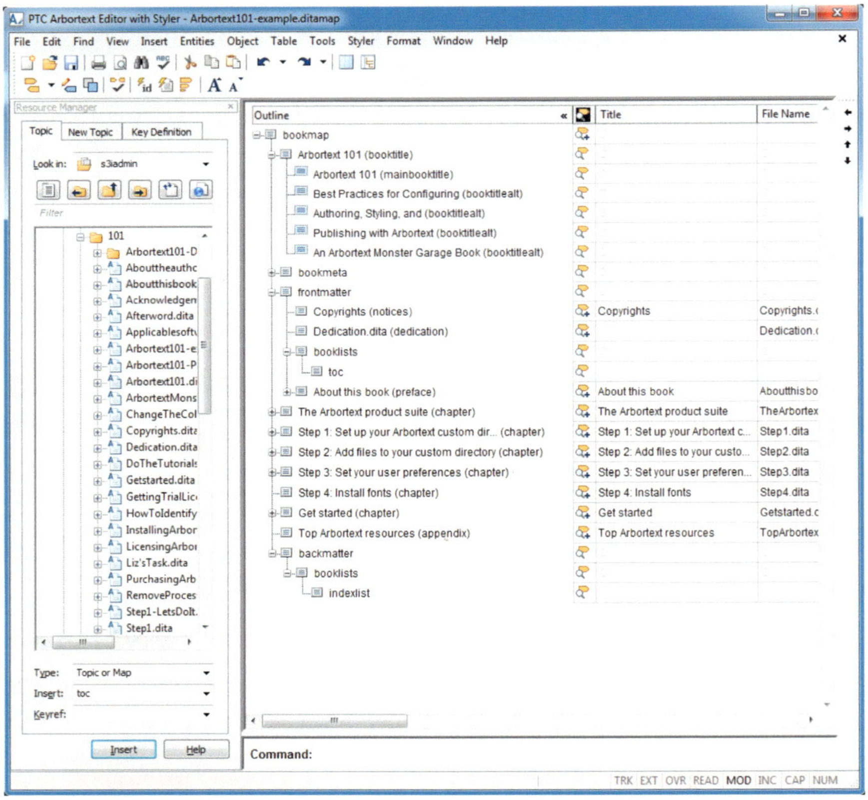

The sample document has at least one of each element that represents each major document division: **preface**, **toc**, **chapter**, **appendix**, and **indexlist**. You can use the rule for each element to trigger the page set changes. Each element will trigger one of the page sets you have already created to handle each of these divisions.

The page sets change as the processing encounters each division element that represents that section of the book content. When you trigger a page set on an element, you are indicating to the stylesheet when the page set should start and how long it should be in use.

For example, when the rendering engine encounters the **toc** element, no matter what page set is currently active the stylesheet will start applying the **Contents Page Set** to the content contained by the **toc** element. Whether the table of contents needs one page or 10 pages, all of the pages generated by processing the **toc** element will use the **Contents Page Set**. When the rendering engine finishes with the **toc** element, the stylesheet will return to the page set that was active before handing the **toc**.

Configure page handling

You'll see how it works as I walk you through setting it up. Once the setup is complete, you'll understand the basics of page handling. After that, it's a short hop to setting up landscape pages, foldouts, and other special page handling situations.

I'm going to walk you through setting up the triggers for the page sets you've already created before going into a detailed discussion about how processing works. Once you've been through it, you'll have a better understanding of how everything works.

This is how you are going to map elements, page sets, and starting page behaviors:

Element name	Page set name	Starting page type
toc	Contents Page Set	First right (odd page)
preface	Preface Page Set	First right (odd page)
chapter	Chapter Page Set	First right (odd page)
appendix	Appendix Page Set	First right (odd page)
indexlist	Index Page Set	First right (odd page)

Configure contents page set execution

The **toc** element should trigger the Contents Page Set and start a new odd page.

Locate the **toc** element rule.

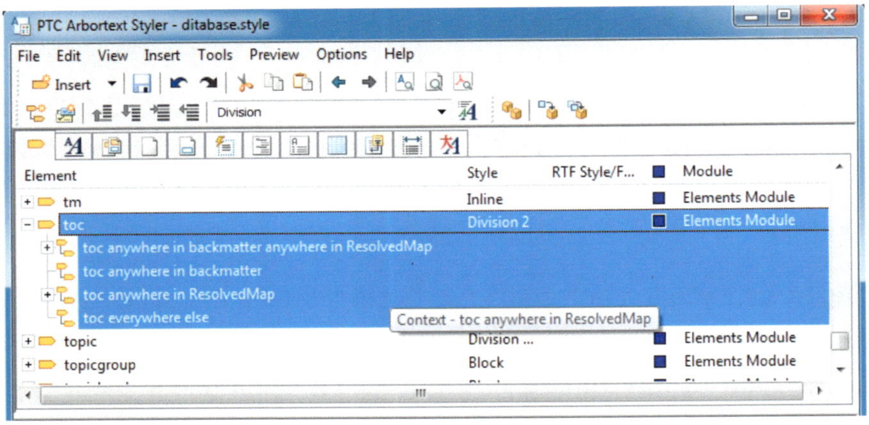

Copy the **toc** element rule to your **My DITA Elements Module**

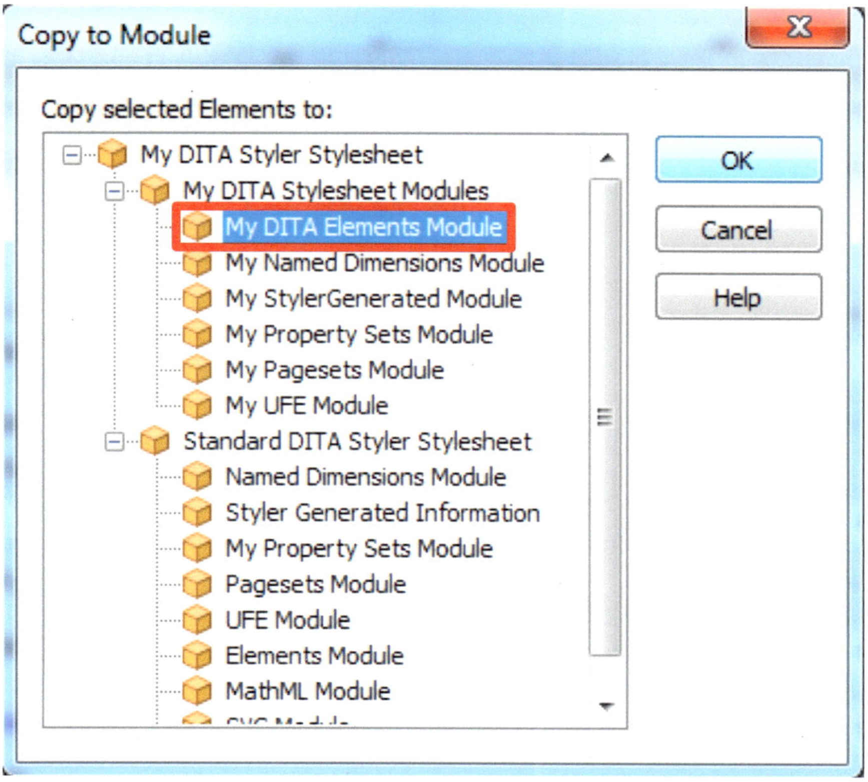

Expand the **toc** rule by clicking the **+** symbol.

The **toc** rule includes four contexts. Two of them have conditions. For now, you don't care about the conditions, but you do care about the contexts. You will need to change two of them.

Here are the contexts and what they mean:

toc anywhere in backmatter in resolved map	In the sample document, the TOC is in the front matter, so this rule does not apply
toc anywhere in backmatter	In the sample document, the TOC is in the front matter, so this rule does not apply

toc anywhere in ResolvedMap	This rule applies. A TOC in the front matter is "anywhere" and every document The Resolved Map is part of the processing pipeline.
toc everywhere else	This rule applies, a TOC in the front matter is included in "everywhere else"

I'll come back to the **ResolvedMap** when I get to the troubleshooting chapter. For now, know that you need to correct the last two conditions.

Select the element rule for **toc everywhere else**.

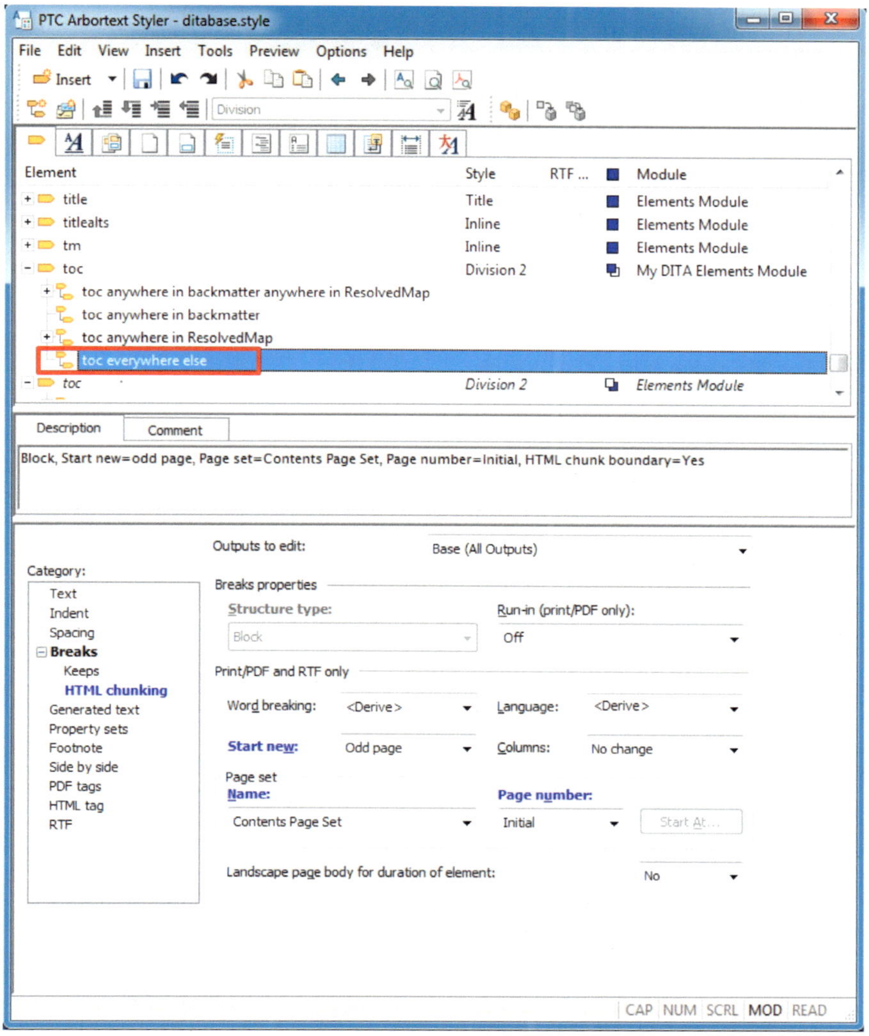

Choose the **Breaks** category, then change these properties:

- **Start new**: Odd page
- **Page set Name**: Contents Page Set
- **Page number**: Initial

Select the element rule for **toc anywhere in ResolvedMap**

Choose the **Breaks** category, then change these properties:

- **Start new**: Odd page
- **Page set Name**: Contents Page Set
- **Page number**: Initial

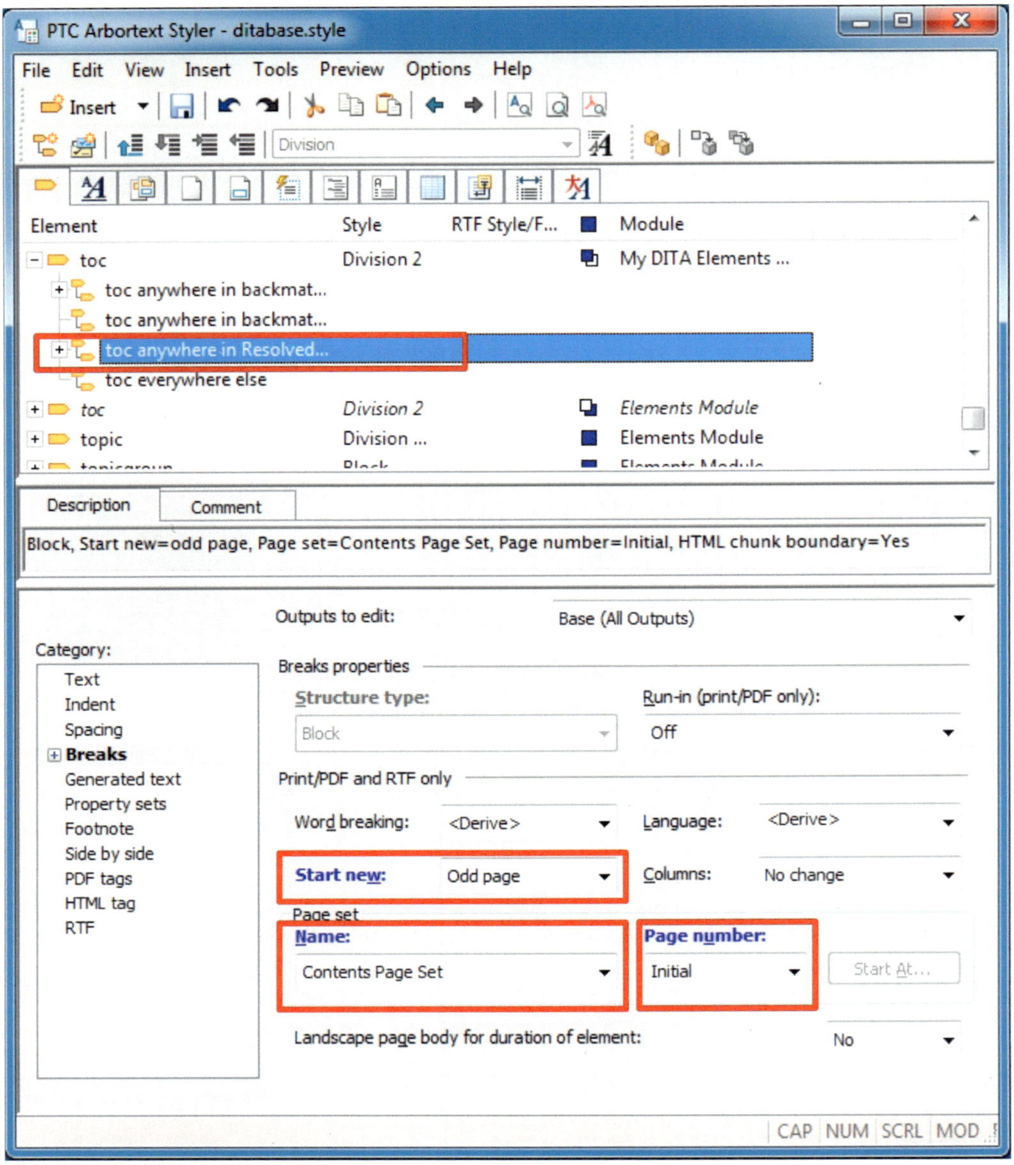

Expert tip!

Always copy a rule to your module before you change it. That way, you preserve the default handling and guarantee that your changes are portable and safe from changes at update time.

Configure preface page set execution

Like the contents page set, the preface element should trigger the Preface Page Set and start a new odd page.

Look for the **preface** element in the element list. It's not there!

The OOTB stylesheet does not include a rule for the **preface** element. The default stylesheet serves double duty: as a starting point for your stylesheets, and as an example of one possible way to set up your stylesheet. The OOTB stylesheet doesn't have a preface element rule because not everyone uses it, and it wasn't a focus for PTC's example stylesheets. That's OK; not everyone uses the preface element. You can add an element rule for the **preface** element to your stylesheet yourself!

Click **Insert**. When you are working in the **Element** tab, it has an **Insert Element** label:

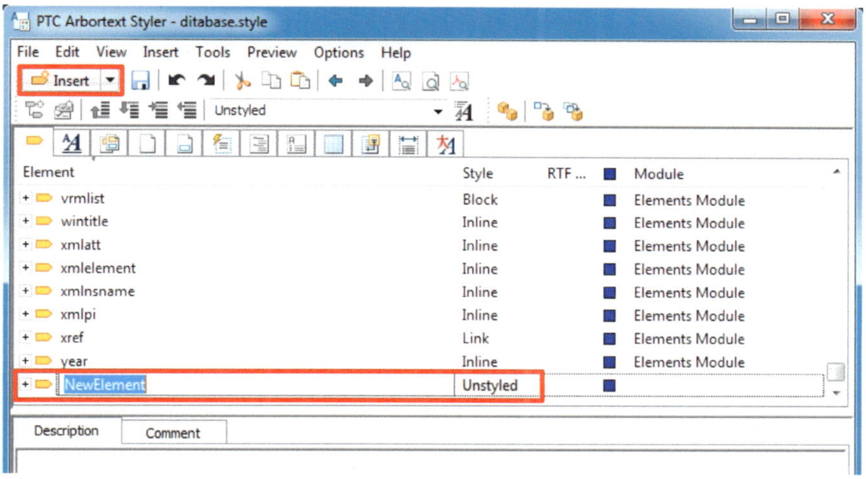

You will get a new rule that simply says **NewElement**. Change it to say **preface**. The element name must be all lowercase and must match the formal element name defined in the DITA doctype.

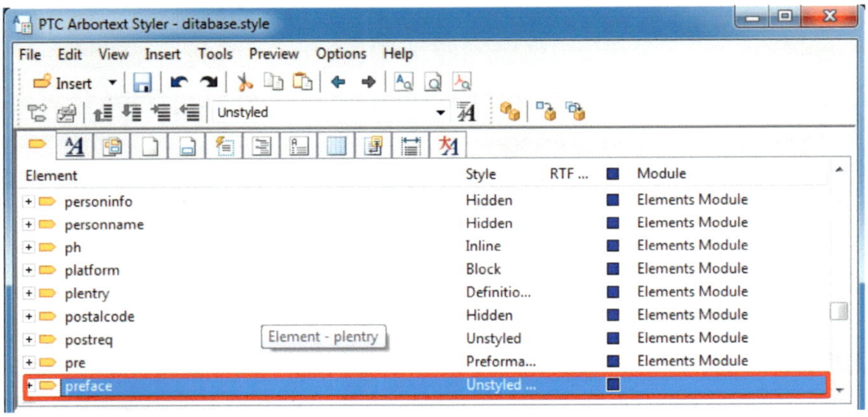

If you want the preface to act as a formal partition of the document (have the topic assigned to it as equivalent to a chapter), change its base style type to **Division**.

Right-click on **preface** element rule and choose **Style→Division**.

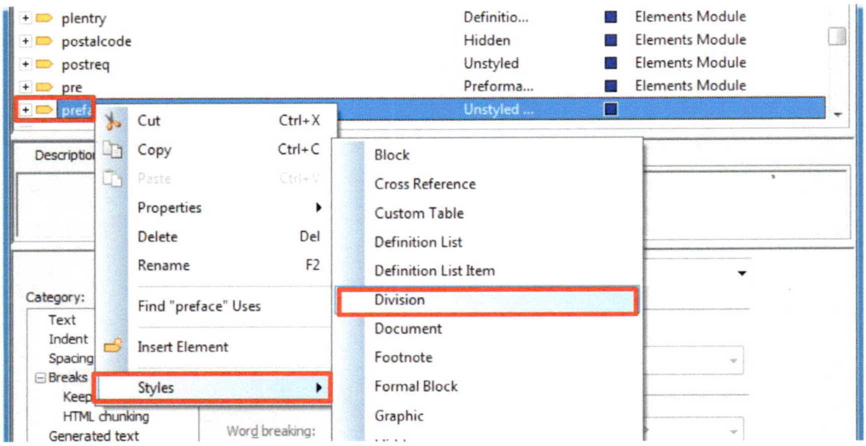

When you choose the **Division** style type, the stylesheet requires you to provide additional information with regard to the hierarchy the new division element holds in the greater context. In this case, you want to set the preface equal in hierarchy to chapters.

Set the **Division Level** to **1** and click **OK**.

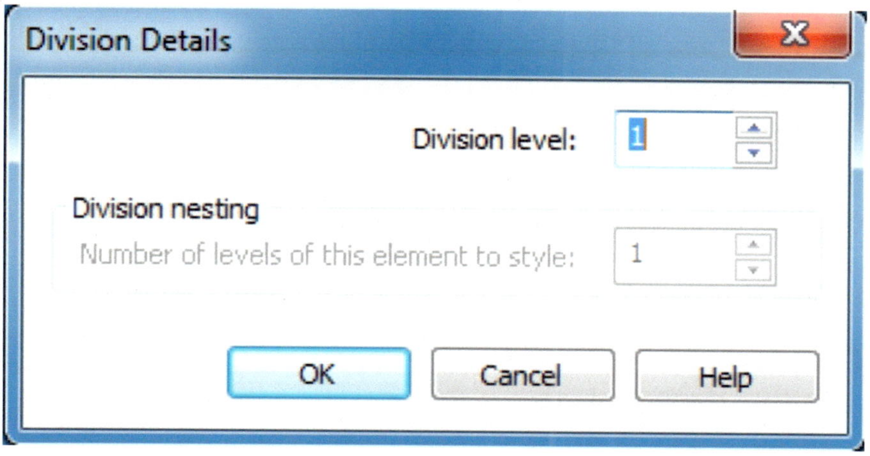

Before you're done, be sure to move the new rule you created to your **My DITA Elements Module**, so that this rule is stored safely in your stylesheet modules.

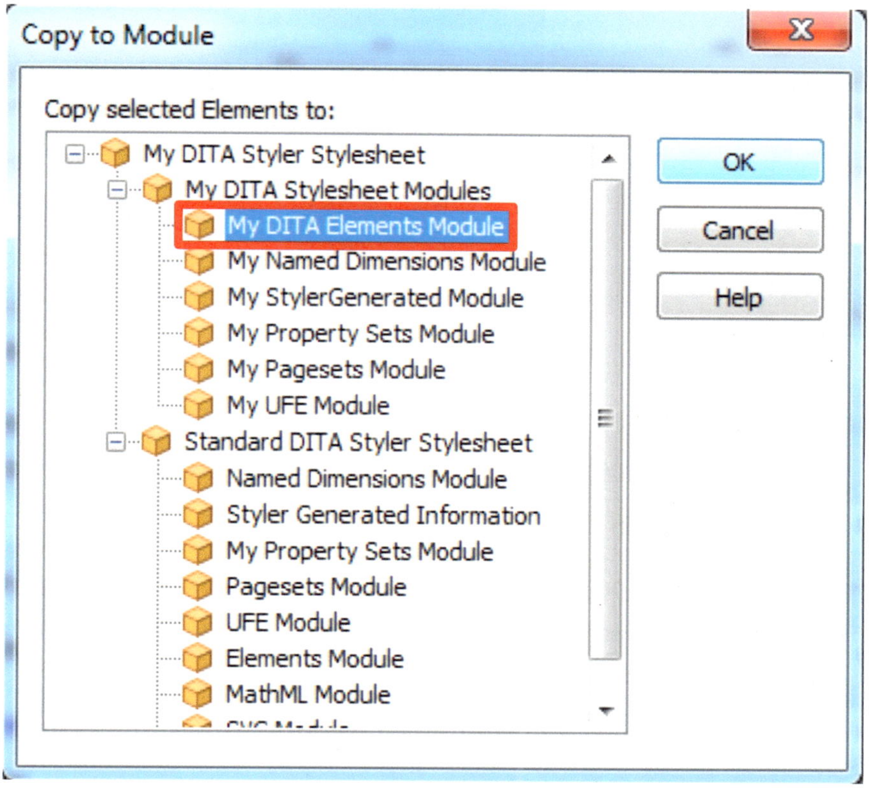

Once you have the rule created and assigned to the proper module, you can apply page-handling properties to the rule.

Select the **preface** rule.

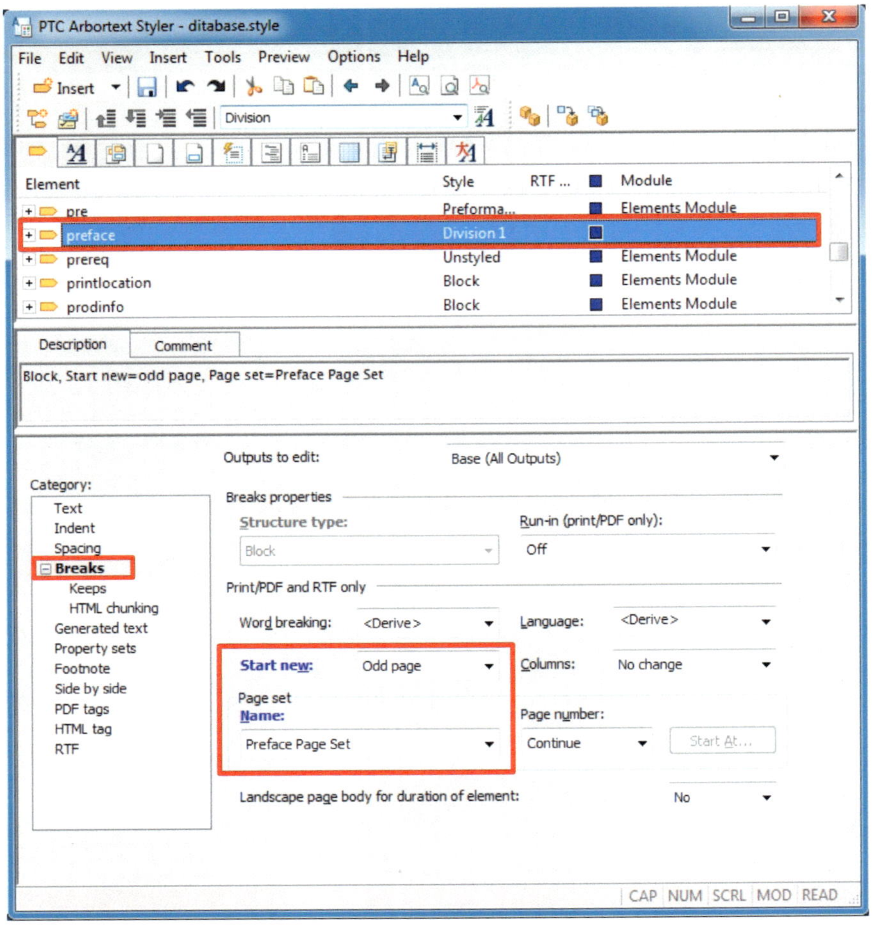

Choose the **Breaks** category, then change these properties:

- **Start new**: Odd page
- **Page set Name**: Preface Page Set
- **Page number**: Continue

This time, choose **Continue** for page numbering. You want the preface to continue numbering the pages. Page numbering (in Roman numerals) started with the Table of Contents. If you don't want to restart numbering with page **i** again when the preface division begins, be sure to set page numbering to **Continue**.

Configure chapter page set execution

Chapter elements should use the chapter page set and they should start on a new odd page. The first chapter should start renumbering using Arabic numeral page numbers. The second and all subsequent chapters should continue page numbering where the previous chapter left off.

Locate the **chapter** element rule and copy it to your **My DITA Elements Module**. Like the **toc** element rule, the **chapter** rule has several contexts already:

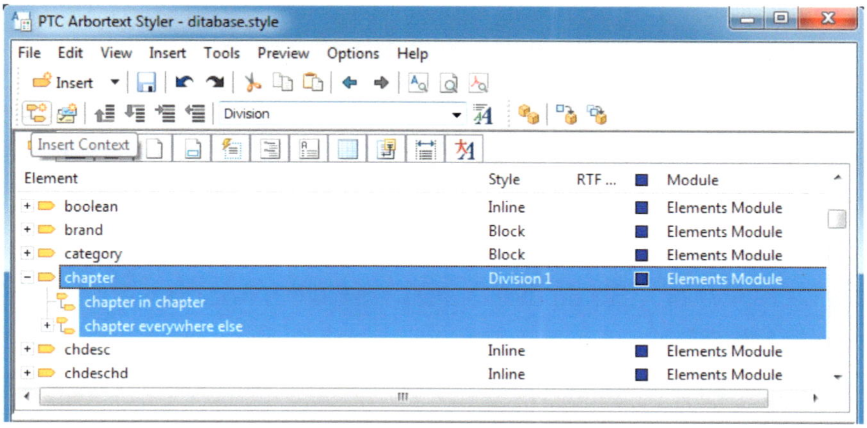

Here are the contexts and what they mean:

Context	What it means
chapter in chapter	In the sample document, I do not use this structure. You can ignore it.
chapter everywhere else	This rule applies to all chapters that are **everywhere else**

The **chapter everywhere else** context has conditions, but you can ignore them for now. However, you want to add a new context for the **first chapter** so you can restart the page numbering when the processor encounters the first chapter. Then, all other chapters fall into the **chapter everywhere else** context and all of those can continue the page numbering.

If you take a closer look at **chapter everywhere else** context rule, you will see that it's already set up to use the **Chapter Page Set**, to start an **Odd Page**, and **Continue** page numbering. All you have to do is add a new context for the **first chapter** and configure it accordingly.

Select the **chapter** element rule (the rule and all its contexts will automatically be selected too) and then click **Insert Context**.

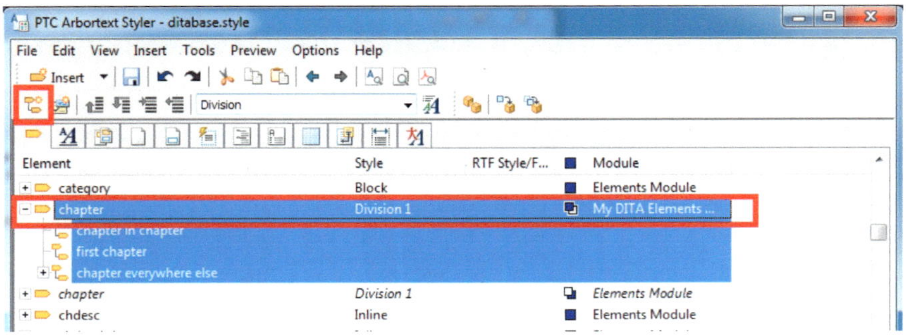

In the **New Context** window, you want to specify that this is a special context for the first chapter only. To do that, select **first** in the **Position** dropdown menu, and click **OK**.

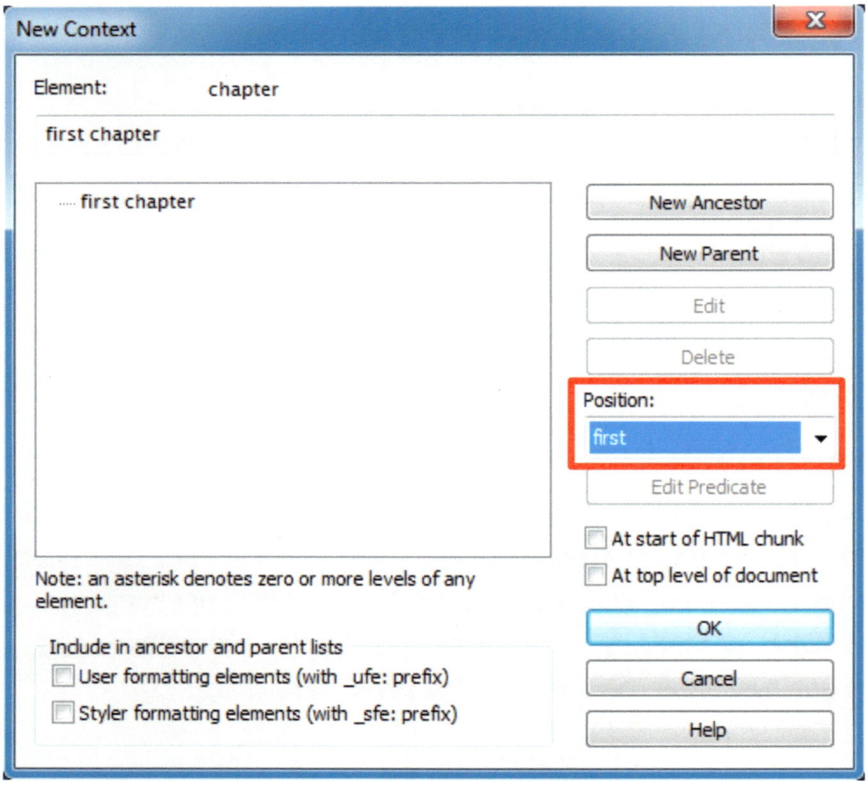

The new context is added to your stylesheet automatically.

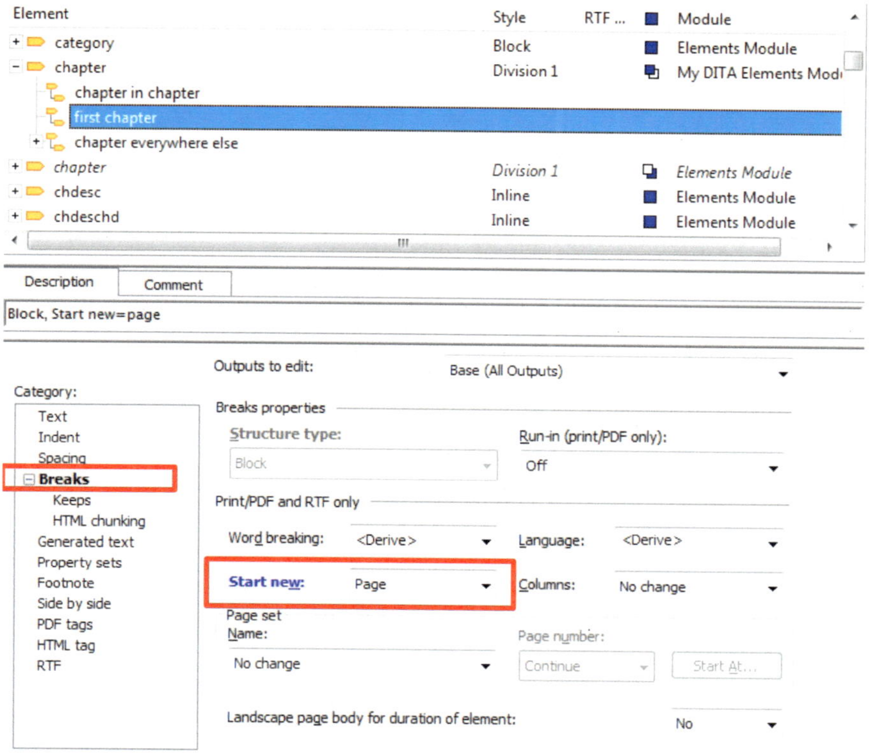

Select the new **first chapter** context so you can apply the proper settings to it.

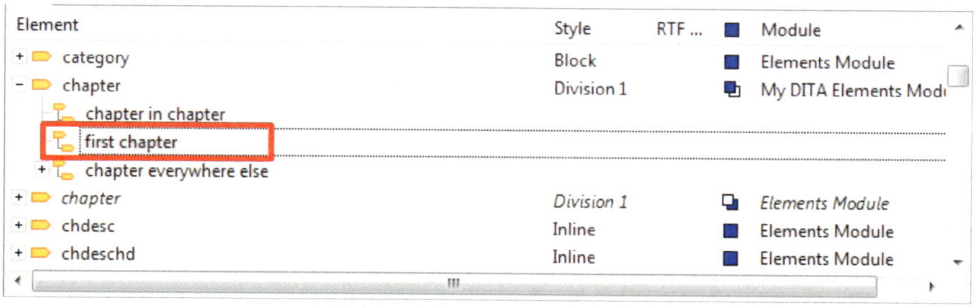

Choose the **Breaks** category, then change these properties:

- **Start new**: Odd page
- **Page set Name**: Chapter Page Set
- **Page number**: Initial

Save your stylesheet.

Configure appendix page set execution

Find the **appendix** element rule. This rule is very simple—it only has one context: **appendix everywhere**.

Copy the **appendix** element rule to your **My DITA Elements Module**.

With an appendix, you don't need to restart the page numbering, so you don't need to add a context like you did for chapter. All you need to do is to change to appendix page set.

Choose the **Breaks** category, then change these properties:

- **Start new**: Odd page
- **Page set Name**: Appendix Page Set

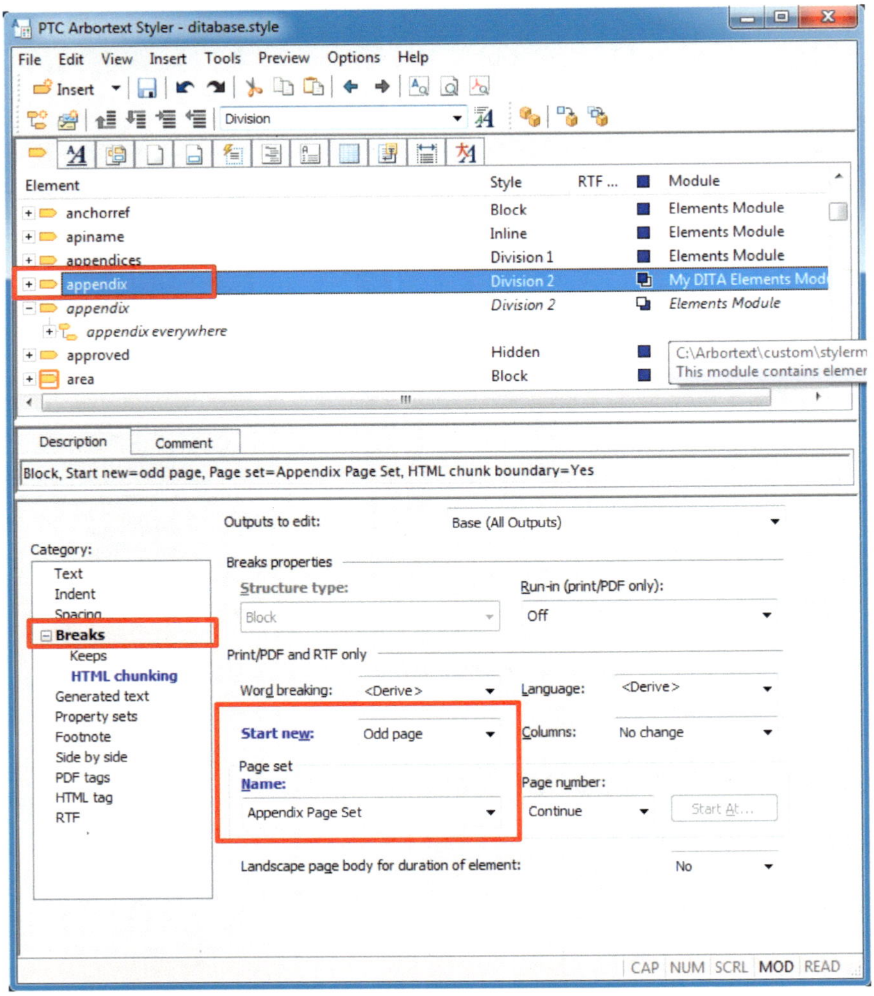

You don't need to change the page numbering property because it is set to continue by default.

Save your stylesheet.

Configure index page set execution

Locate the **indexlist** element rule and copy it to your **My DITA Elements Module**.

Choose the **Breaks** category, then change these properties:

- **Start new**: Odd page
- **Page set Name**: Index Page Set

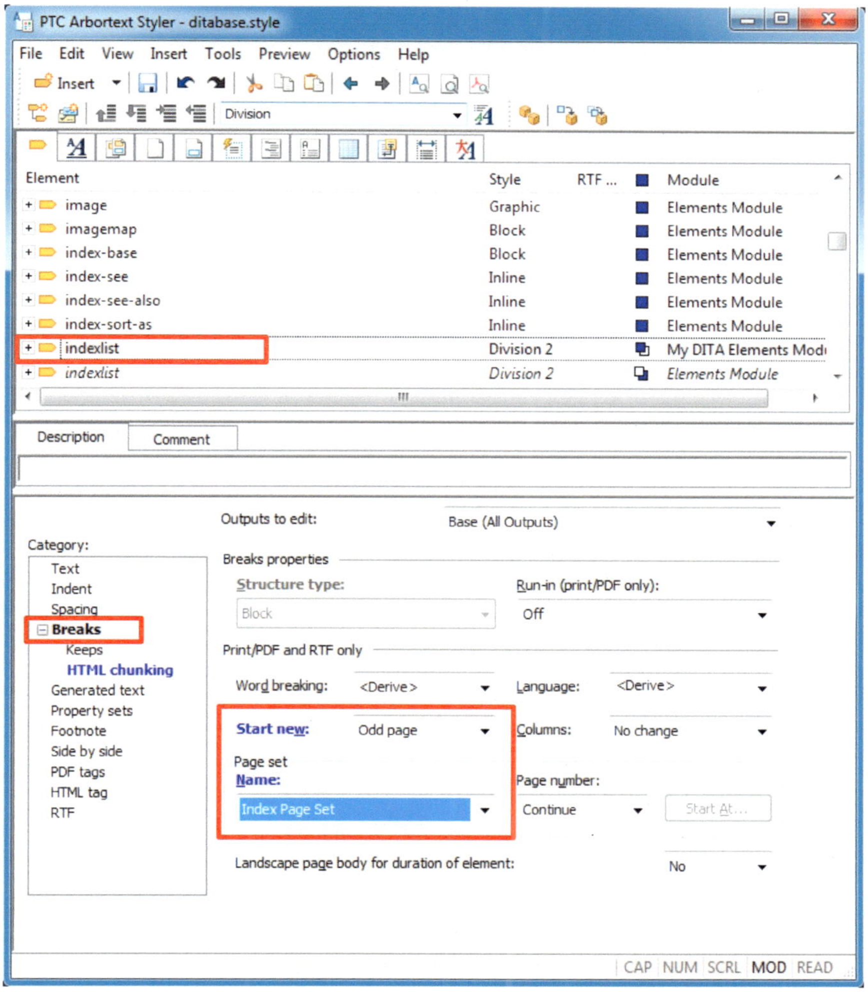

Like the appendix, you don't need to change the page numbering property because it is set to continue by default.

Save your stylesheet.

Chapter 6. Test your framework

At this point you should test your framework. All page sets are configured and are attached to an element rule. If you publish a PDF now using the sample content, you should be able to see your page sets trigger exactly as you expect.

This is a good time to test your framework to make sure the big pieces are in place. It's a milestone moment. Page sets happen as expected. When a PDF is generated, all contents end up on a page. Page numbering is sequential and starts or continues, as it should. The header and footer content displays the correct content based on the configurations you made earlier.

Press the **PDF** button in the stylesheet editor window. Arbortext Styler will start building a PDF.

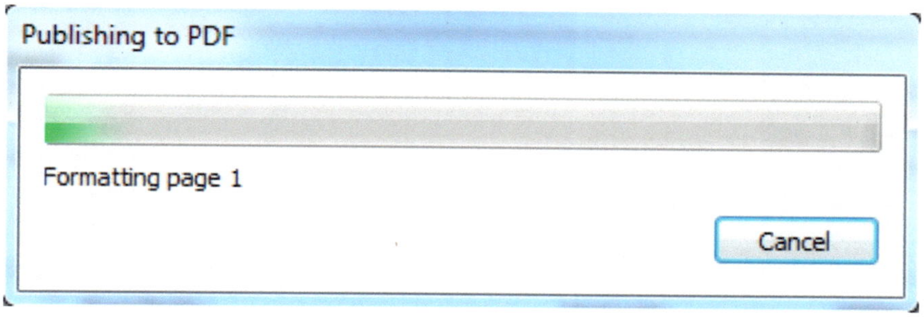

When the PDF appears, forget about the pages before the TOC. Remember, we're handling the cover and related pages last. Page down until you see your TOC and take a look.

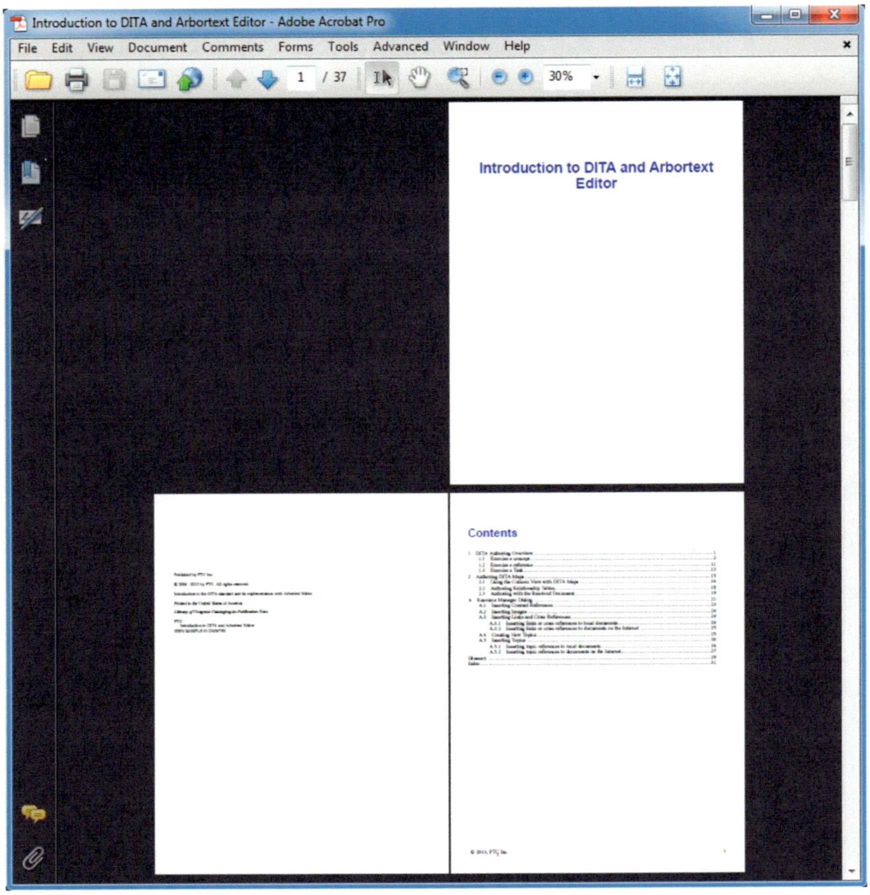

The Table of Contents is starting on a right-hand page. The page numbering starts with Roman numeral **i**. The footer has the copyright date and the company name from the **bookmeta** information in the source document.

We don't have a left-hand page for the TOC, but that's because we don't have enough content in our sample to have the TOC run to multiple pages. You can adjust the sample content as much as you want until you can force the TOC to run to multiple pages. For now, let's move on.

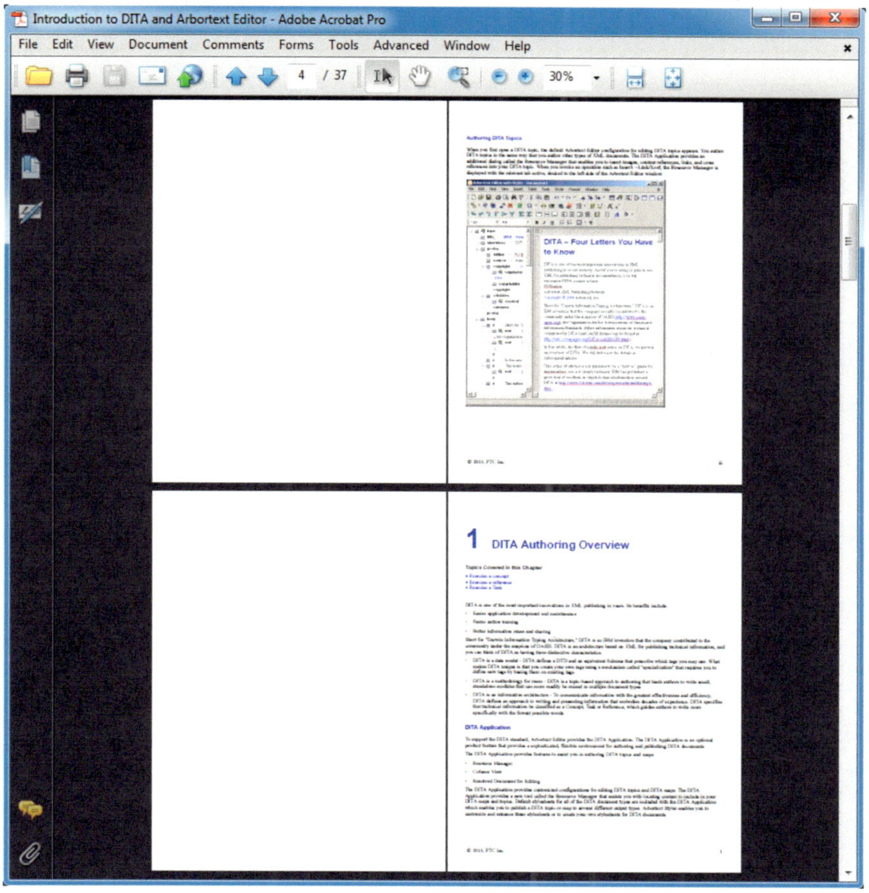

Like the TOC, the preface is a single page, but we can see that it starts on a right-hand page and that the page numbering has continued correctly using Roman numerals.

You can also see that the Chapter—the first chapter—starts on a right-hand page and the page numbering restarts and uses Arabic numerals. Check the footers and you'll see that they also look as expected.

Page forward so you can see other pages of the chapter (second and later pages).

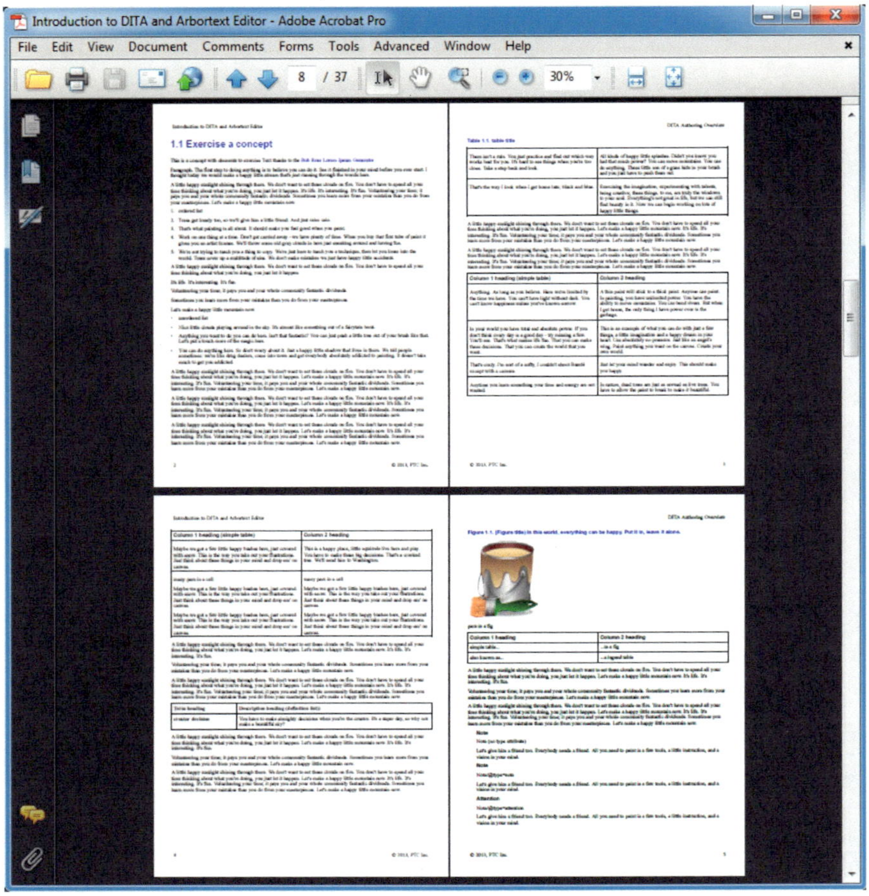

The chapter content runs multiple pages and you can see that the left-hand page mirrors the right. Page numbers are in the outside corners. Header content also is aligned to the outside page edge. You can see the book title in the header of the left-hand page and the running header on the right-hand page.

Page forward until you can see the appendix.

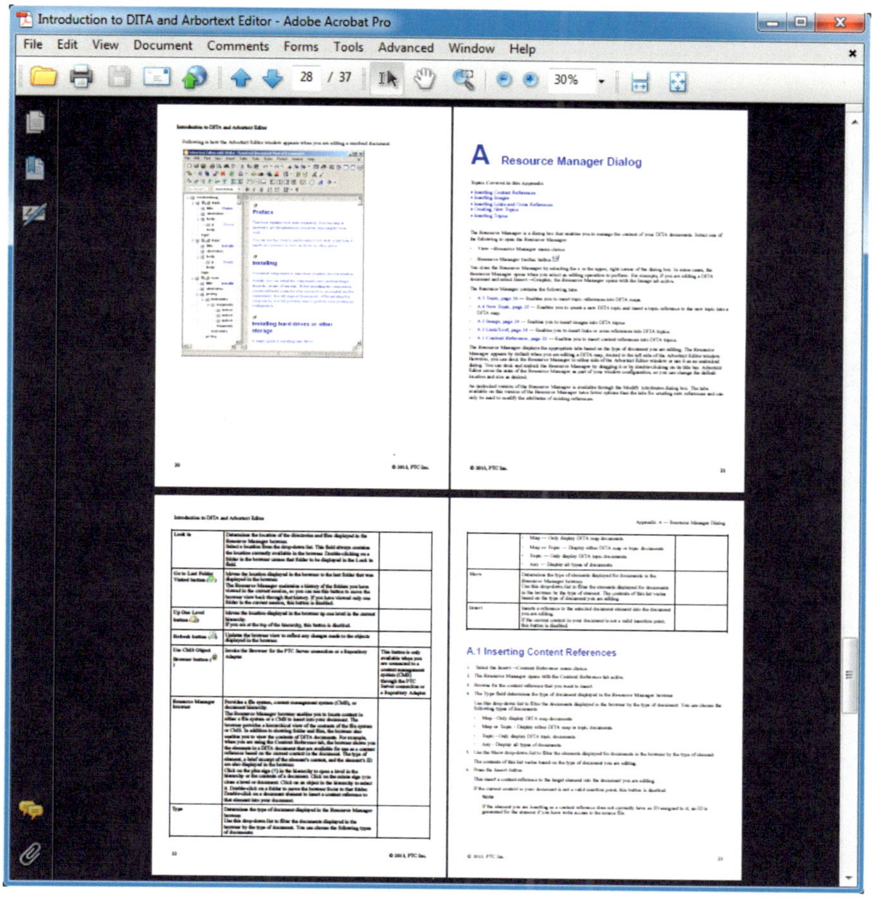

The appendix changes in the header content of the right-hand page, but the page numbering continues, as it should.

Page forward until you can see the Index.

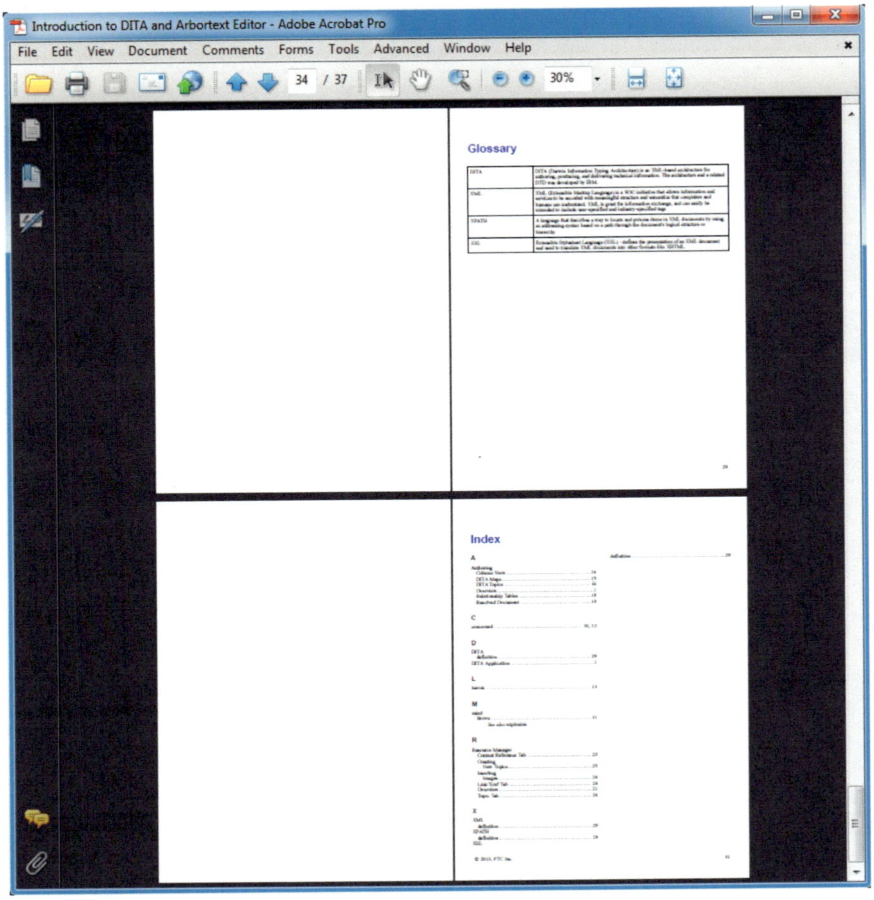

The Index page, like the TOC, has only enough content to fill a single page, but it also looks correct. Page numbering continues and it starts on a new right-hand page.

Your framework is complete. Your stylesheet is laying your content on the page. It's displaying typical behavior for page layout and executing page layouts as appropriate to the content in your sample document.

The heavy lifting is done. Now you can adjust the rest of the element rules and remove formatting properties like the numbering on chapters or changing the generated text on chapter titles so they say **Chapter 1** (instead of just **1**).

You can adjust the colors, fonts, and anything else you want.

You can also start working on the cover. The current stylesheet does give you a cover and an inside front cover page. You can now use the lessons you learned in this chapter to create a shiny new cover that fits your business requirements.

Expert tip!

If something isn't right, verify that all the page sets, page types, page regions, and page contents match. Because we named everything consistently—chapter page set has chapter page types, which have chapter page regions, which have chapter generated text content—you can verify your settings easily.

Always make sure your page sets are working before starting on any other formatting tasks.

If you still have a problem, see *Chapter 12. the debugging chapter, page 245*.

Chapter 7. Property sets

Now that your framework is complete and you have content appearing on the page, you can change the formatting properties of individual elements and know that there won't be any unexpected side effects.

If you start with the element rules before you lay the framework (your page sets, page types, page regions and generated text), you can easily get distracted and end up with more work in the end. You can't hang a curtain if the window isn't already in the wall. If you do, you might catch the curtain when you install the window and end up with the curtain passing through the wall underneath the window frame.

Likewise, you can't fully understand how to add formatting properties to an element rule effectively until you understand how inheritance affects processing and how property sets work.

All XML processing works in the same way. The document is fed into the rendering engine one element at a time. The top element (**bookmap** in our sample document) is sent first. Then its first child, then the child's first child, and on and on until there are no more elements; then it moves to the second child and repeats the process. It does this recursively until all elements have been processed. The rendering engine looks beyond document boundaries and processes all elements of external files as if they appear in place.

Because the parent element is sent to the rendering engine before its child elements, you can add formatting properties to the parent that the children inherit.

For example, the OOTB stylesheet uses a standard Serif 11pt font for print output. Open your sample document and your stylesheet. Place the cursor in the **bookmap** element in the sample document, then look at the stylesheet editor. It will automatically sync to the element you selected in the document editor. In this case, the stylesheet editor should have the **bookmap** element rule selected:

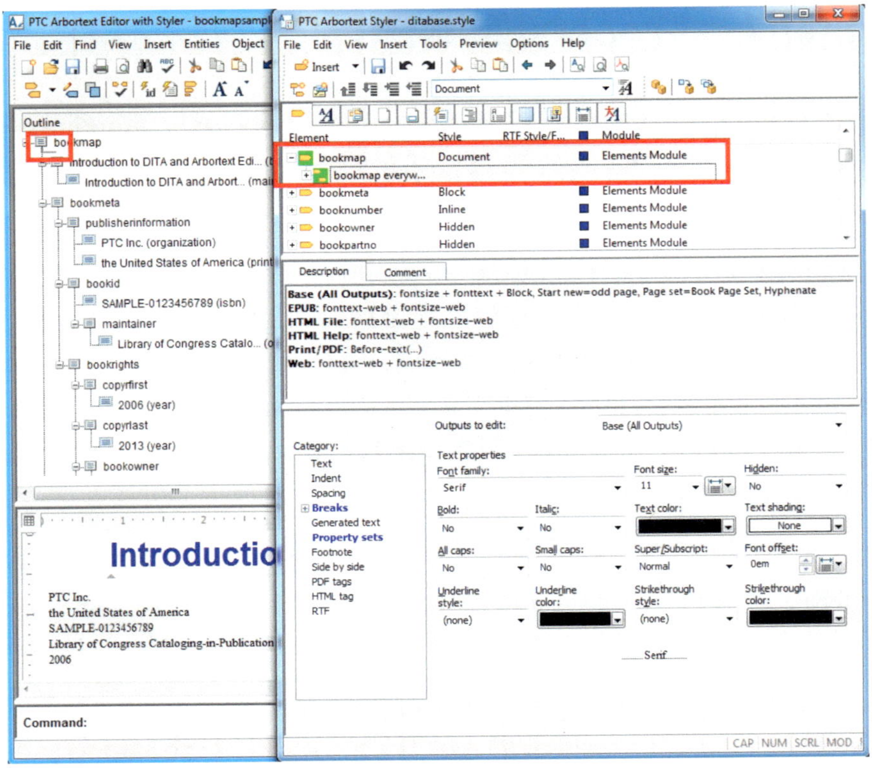

If you look at the lower panel, you will notice that some categories are blue and some are black. If you see blue highlight used in the UI, it means that particular property has settings in this element rule.

For **bookmap**, only the **Breaks** and **Property sets** categories are blue. However, the **Text** category has values set. It says Serif font, 11 pt.

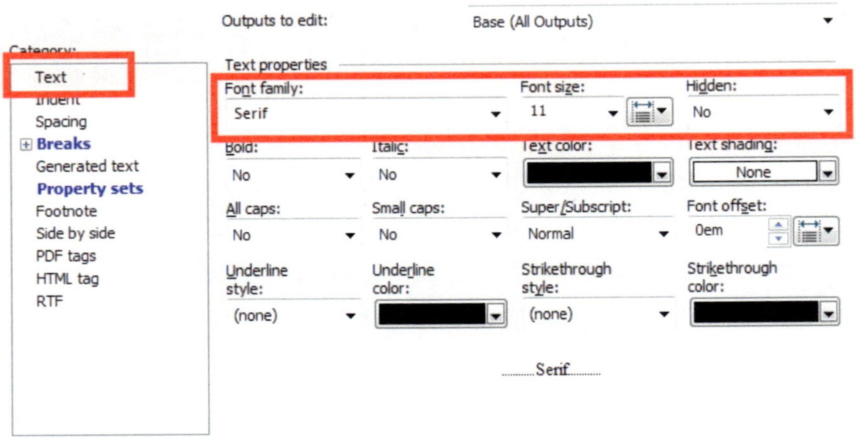

When a property has a value but the blue color is absent on the property's name, the property is inheriting that value from somewhere else in the stylesheet or another setting elsewhere.

In this case, notice that the **Property sets** category is blue. That's always the first place to check. When you click the **Property sets** category, you see that two property sets are applied to this rule: **fontsize** and **fonttext**. (Also, note that the **Used property sets** UI text is blue.)

Click **fontsize** then click **Go To** .

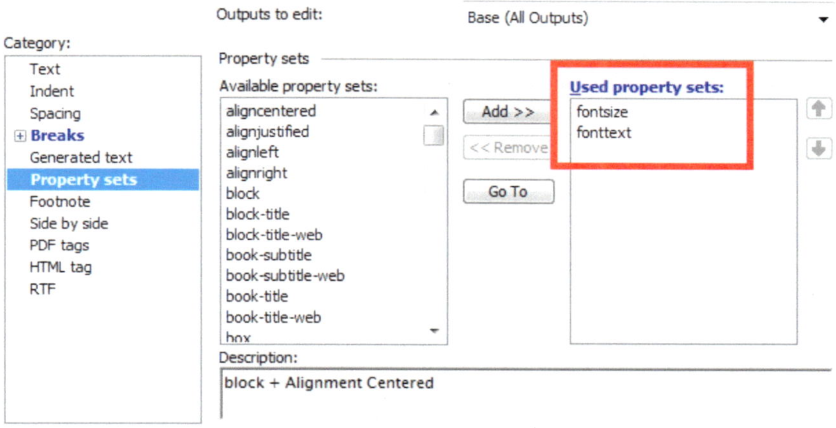

This action will change the UI to display the definition of the **fontsize** property set, which is in the **Property sets** tab.

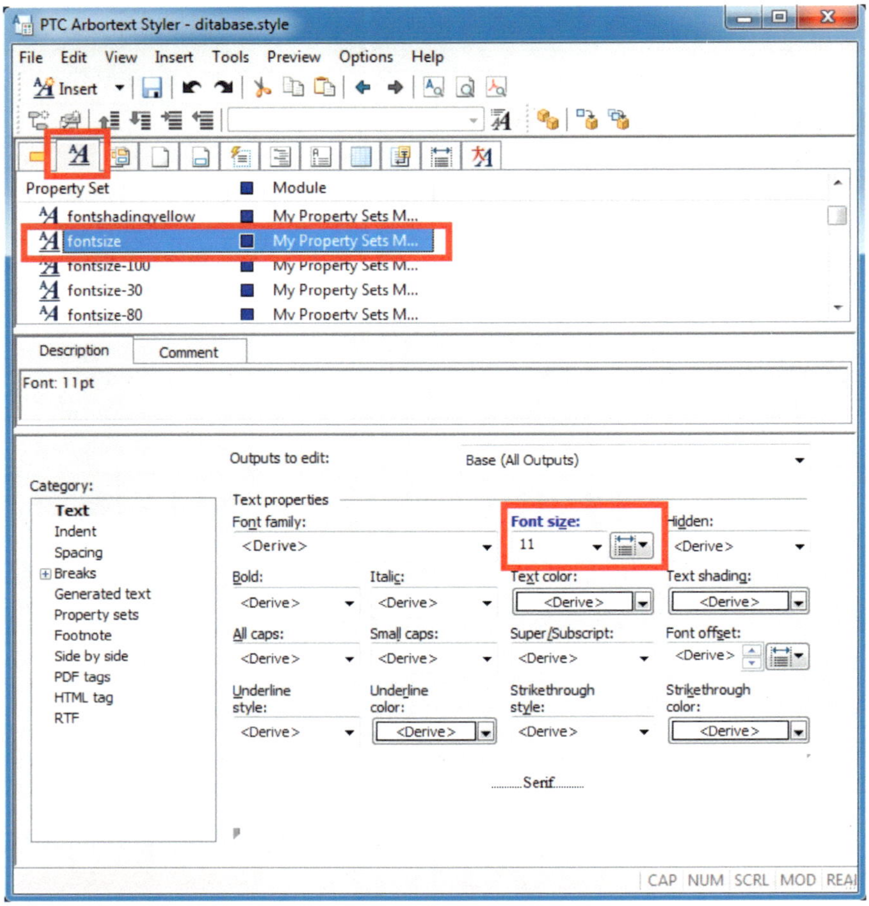

You can see that the **fontsize** property set has the **Font size** formatting property set. It says 11pt. Because the **bookmap** element rule has the **fontsize** property set applied to it, it inherits the 11pt specified in that property set.

By using property sets, you can decouple formatting-specific properties from the element rules and then reuse the property sets over and over on as many element rules you want. Then, if you want to change the default font size from 11pt to 14pt, you can change this one property set. Any element that inherits its value will automatically switch to 14pt.

Where have you heard this before? It's *reuse*, applied within a stylesheet. You're creating formatting property libraries that you can use over and over. Just like you create topics you can reuse, with Arbortext Styler you can create property sets that you can reuse. By isolating formatting into a property set, you get the power of reuse where you need it most.

No more tracking down how many places you set the font to 11pt. What a nightmare that would be! How many elements are in your doctype? How many contexts? You don't want to have to specify 11pt font for each one of them. Instead, you create one property set, add it to all the rules *that need it*, and then you can change its definition, everywhere all at once! It's that same power of XML that you use for your content—applied to a stylesheet!

When you add the other inheritance property, you simplify your task even further. The **bookmap** is the ancestor element of every other element in your document. If you set the font size in its element rule, every element under it will inherit that value. That means that you have to include the **fontsize** property set only in this rule. All other child elements (and their child elements) will inherit the 11pt font size you set here.

Inheritance is powerful. You can inherit from a parent element (or grandparent or other ancestor), *and* you can inherit from specific property sets that you add to an element rule.

But you might wonder: How then do you get other font sizes? You add a different property set to the element rule you want to be different!

For example, look at **mainbooktitle**. In the document editor, place the cursor in the **mainbooktitle** element. Then, click the **Elements** tab in the stylesheet editor to look at the **mainbooktitle** element rule.

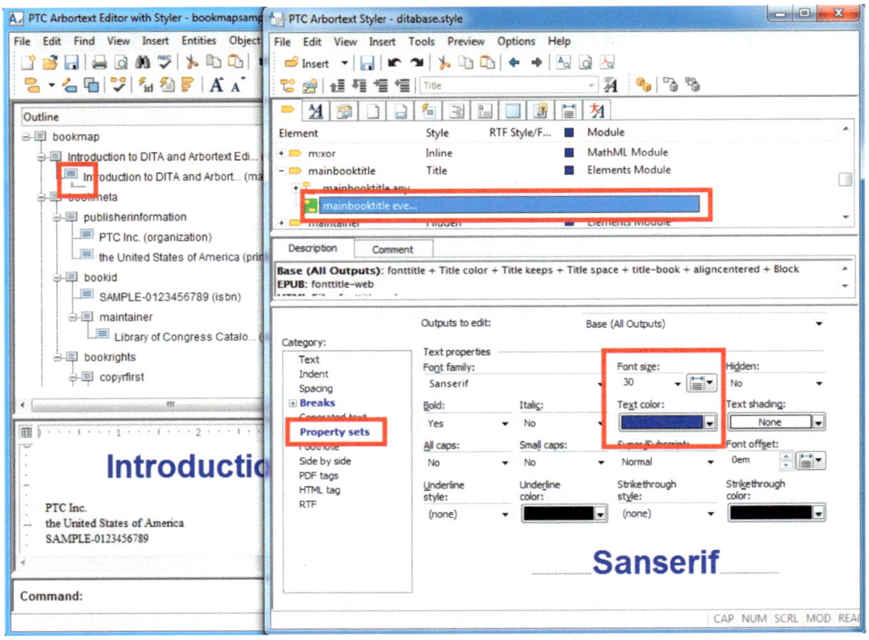

Although the **Text** category is not blue, you can see that this rule has different settings for the **Font size** and the **Text color**. You want to check the **Property sets**: It's blue.

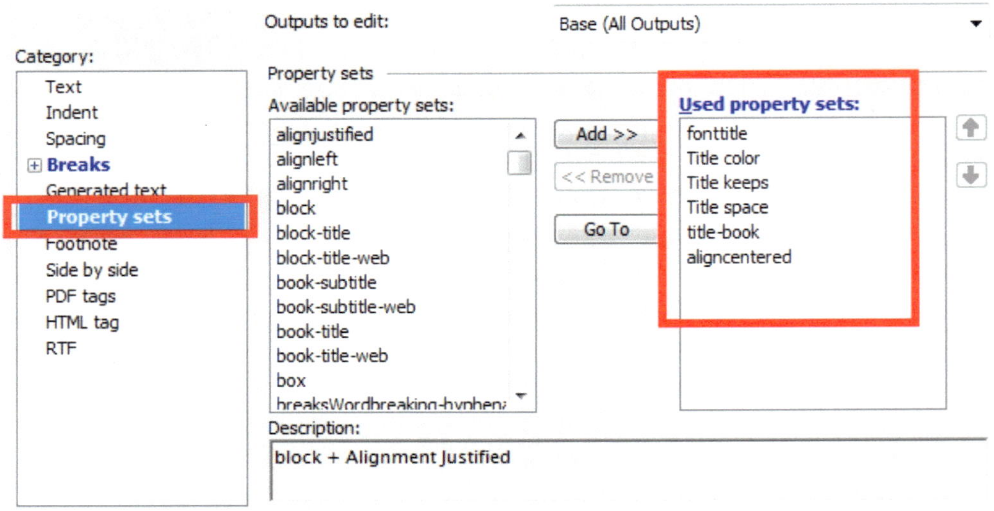

This rule has a lot more than the two property sets that the **bookmap** rule had. It's got six property sets:

- **fonttitle**
- **Title color**

- **Title keeps**
- **Title space**
- **Title-book**
- **aligncentered**

You can select each one and then click **Go To** to see how each one is defined. Even though the **mainbooktitle** is inheriting from **bookmap**, settings on the individual rule for **mainbooktitle** override the settings that it inherits. Here again, no settings are in the **Text** or **Indent** categories—neither one is blue. This rule, like the other, depends on property sets to add formatting properties to this rule.

Wherever possible, use property sets instead of setting individual properties on an element rule. Property sets give you the same power to specify specific formatting while providing more flexibility and the ability to set it once, change it everywhere! Changes to look and feel will come eventually, typically when you have the least time to spend on fixing the stylesheet. By using property sets you won't have to track down individual settings. You'll thank yourself later. Trust me.

Next to modules, property sets are the most powerful tools in your Arbortext Styler toolbox.

Expert tip!

Always define property sets that you create and modify in your own property set module. Property sets are element neutral. They represent formatting properties only. You add a property set to an element rule to apply formatting properties to that rule. This means that a property set module is truly reusable. By putting your property set rules in your property set module, you can include the module in any stylesheet you create. You can then use them, regardless of the doctype your stylesheet is designed to support.

Chapter 8. Formatting Elements

Topics Covered in this Chapter

- ♦ How to fix the chapter title
- ♦ How to remove numbering from sections
- ♦ How to fix the titles in the preface
- ♦ How to change to multi-column layout
- ♦ How to set base document properties
- ♦ User Formatting Elements (UFEs)

The **Elements** tab displays the list of element rules defined in your stylesheet. Generally, you have one element rule for each element defined in your document type. You define formatting properties on for an element in your source documents by adding the formatting to the representative element rule. When the rendering engine encounters the element in running text, it looks at the matching element rule, and formats the element accordingly.

Determining the correct "match" between an element and an element rule can be a complex task. For example, a chapter title looks different from a figure title. One says "Chapter" and the other says "Figure" before the title text. Both are numbered. A heading 2, the title to a subordinate section in a chapter looks different still. It might or might not be numbered. It might have a slightly smaller font. It might have color, emphasis, or italics applied.

In all three cases, the element is **title**. It is the context—the **title in chapter** vs. **title in figure** vs. **title in section** – that determines the formatting. It is the context that determines the match requirement. Many elements have multiple contexts that affect their formatting. It's up to you, the stylesheet designer, to make sure the contexts are fully defined so the rendering engine can know what to do in each situation. A context describes the situation sufficiently for the rendering engine to be able to distinguish one context from another and make the correct contextual match.

Figuring out which context is the right one can be difficult. It can also bring complexity if you start changing styling properties without thinking through all the possible contexts for the element you want to format.

For example, what happens if you add space after every paragraph? As long as all you have are paragraphs, everything works fine. Every paragraph works the same way. Every paragraph is preceded by or followed by another paragraph. The uniformity works as long as the content matches. Unfortunately, it can give you a bit of tunnel vision.

For example, what happens if you insert a table? Does the paragraph that follows the table run right up against it? Do you add space after tables too? Or, do you add space before paragraphs as well as after?

And what about all the other elements that can appear in running text? Definition lists, notes and warnings, sections, figures, and all the other text elements that can appear in running text? What about the infinite possible combinations of those elements that a writer can construct? You can't count on any specific structure for text. Authors control that when they're writing the content.

Before you get overwhelmed, know that Styler is designed to help you out in these situations. When you incorporate the out-of-the-box stylesheets as modules of our own, most of these situations are already handled. You can make adjustments without having to handle every possible permutation manually.

I'm going to walk you through several examples that represent typical formatting changes:

1. How to fix the chapter title
2. How to remove numbering from sections
3. How to format the preface title so it matches other head 1 formatting (like chapter titles)
4. How to change the index from 2-columns to 1-column layout

I'll explain what I look for and the process I follow when making formatting changes. I'll also provide step-by-step instructions to make the changes listed above. When you're done, you should have a sense of how to make changes to formatting properties with confidence.

How to fix the chapter title

The default stylesheet formats a chapter title like this for print output:

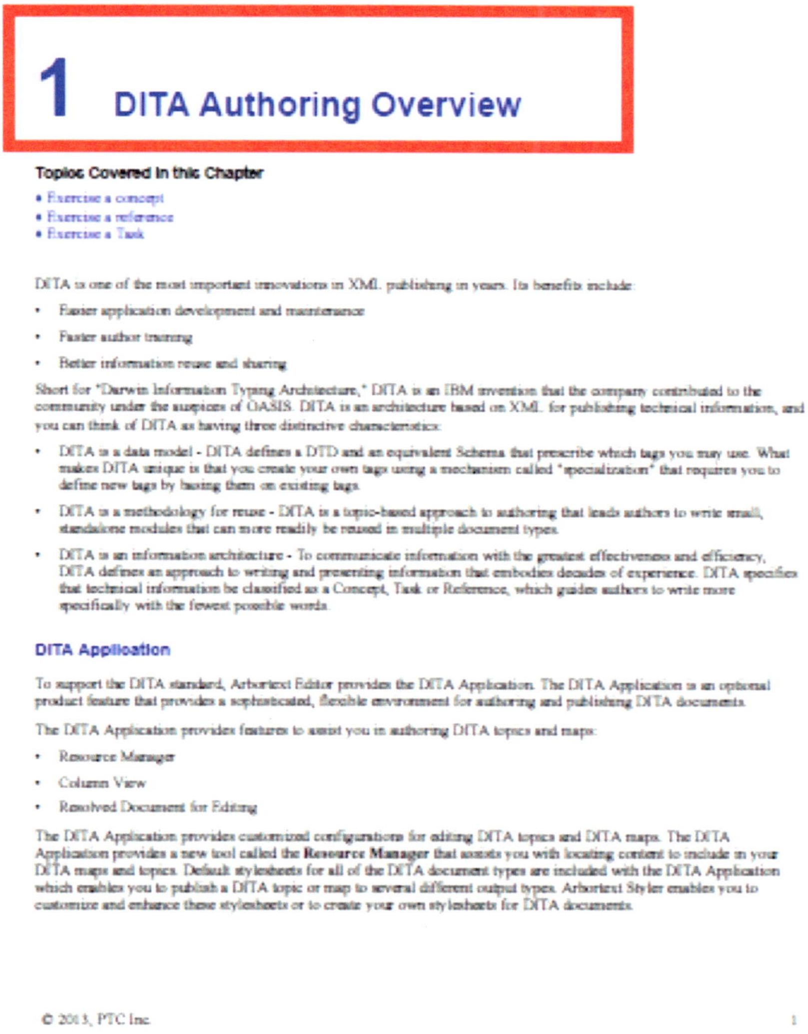

1 DITA Authoring Overview

Topics Covered in this Chapter

- Exercise a concept
- Exercise a reference
- Exercise a Task

DITA is one of the most important innovations in XML publishing in years. Its benefits include:

- Easier application development and maintenance
- Faster author training
- Better information reuse and sharing

Short for "Darwin Information Typing Architecture," DITA is an IBM invention that the company contributed to the community under the auspices of OASIS. DITA is an architecture based on XML for publishing technical information, and you can think of DITA as having three distinctive characteristics:

- DITA is a data model - DITA defines a DTD and an equivalent Schema that prescribe which tags you may use. What makes DITA unique is that you create your own tags using a mechanism called "specialization" that requires you to define new tags by basing them on existing tags.
- DITA is a methodology for reuse - DITA is a topic-based approach to authoring that leads authors to write small, standalone modules that can more readily be reused in multiple document types.
- DITA is an information architecture - To communicate information with the greatest effectiveness and efficiency, DITA defines an approach to writing and presenting information that embodies decades of experience. DITA specifies that technical information be classified as a Concept, Task or Reference, which guides authors to write more specifically with the fewest possible words.

DITA Application

To support the DITA standard, Arbortext Editor provides the DITA Application. The DITA Application is an optional product feature that provides a sophisticated, flexible environment for authoring and publishing DITA documents.

The DITA Application provides features to assist you in authoring DITA topics and maps:

- Resource Manager
- Column View
- Resolved Document for Editing

The DITA Application provides customized configurations for editing DITA topics and DITA maps. The DITA Application provides a new tool called the **Resource Manager** that assists you with locating content to include in your DITA maps and topics. Default stylesheets for all of the DITA document types are included with the DITA Application which enables you to publish a DITA topic or map to several different output types. Arbortext Styler enables you to customize and enhance these stylesheets or to create your own stylesheets for DITA documents.

© 2013, PTC Inc. 1

By default, chapter titles are numbered. The number has a bigger font size than the rest of the title. Only the chapter number appears. The title doesn't have a label (such as "Chapter") before the chapter number. Both the title and the chapter number are blue.

You can change any or all of these properties. You just have to know what to change and where to change it. To figure that out, you need to do some sleuthing because there are element rules and property sets that both contribute to look and feel.

I'll show you how to make two changes:

1. The color (black, not blue)
2. The generated text (content and formatting)
    a. The label should say "Chapter 1" (not just "1")
    b. The size of the generated text should be the same size as the rest of the title

Change the color

Open your sample document and your stylesheet. In the stylesheet editor, click on the **Elements** tab and locate the **title** rule.

You want the **title in chapter** context. Click the plus icon to open up the title contexts and locate the **title in chapter**. It could take you a while to locate the right context: The title rule has 51 context rules.

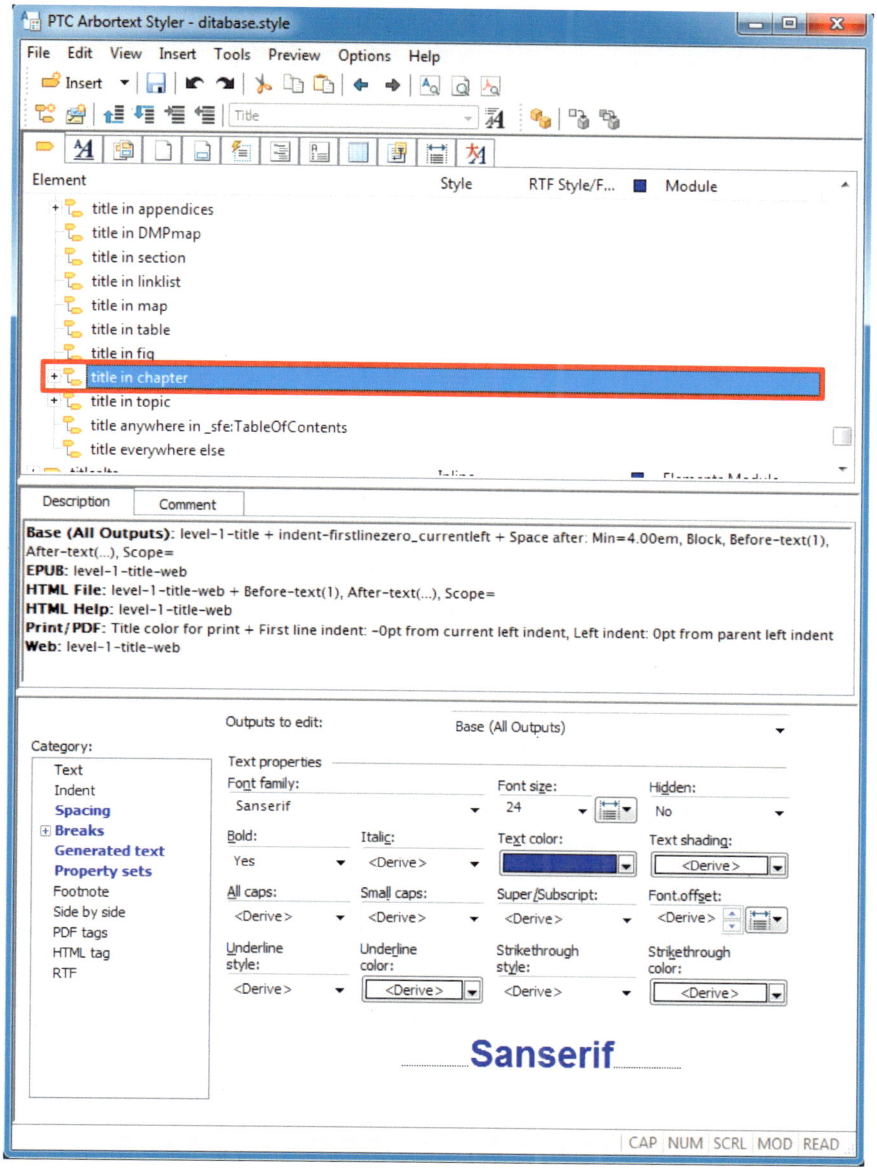

Copy the title element rule to your My DITA Elements Module.

Before you start changing things in this rule, it's important to make sure you are working on the right output. The **title in chapter** context rule has output format-specific settings. You can see them all listed in the **Description**.

Because you want to change the formatting of the title in print output, change the **Outputs to edit** dropdown menu to **Print/PDF**.

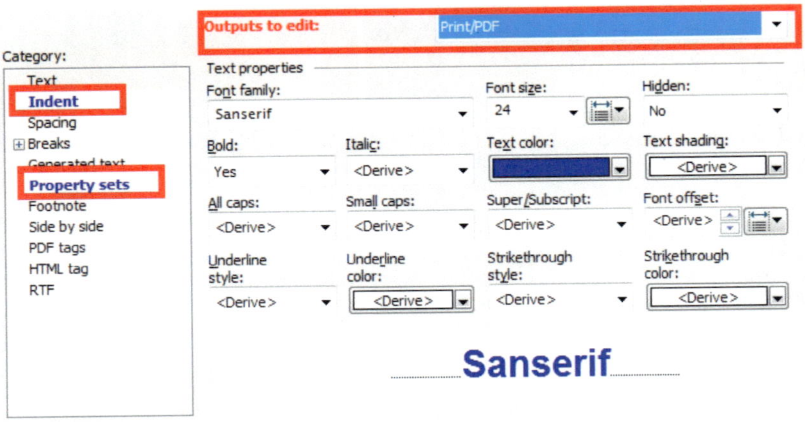

Again, before you make any changes, you want to find out where the current settings are. That means looking at any property names that are **blue**. A blue property name means that specific settings are applied to that property.

In this context rule, for the Print/PDF output, only two categories are blue:

- Indent
- Property sets

Click **Indent** in the **Category** pane.

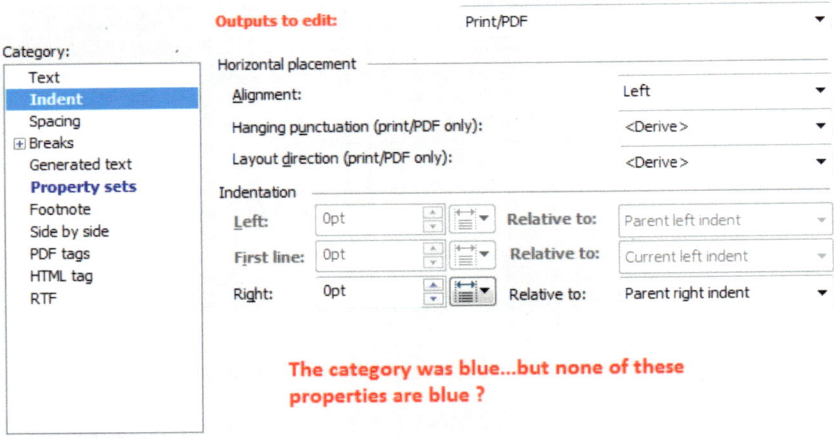

The category was blue but none of the properties in this category are blue. This means that there are settings somewhere else that are affecting indent properties.

Click **Property sets** in the **Category** pane.

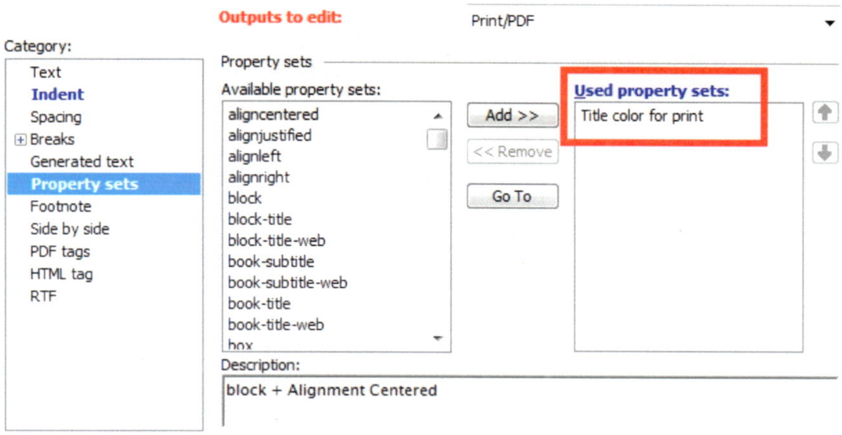

Dig into that property. Select **Title color for print**, then click **Go To**.

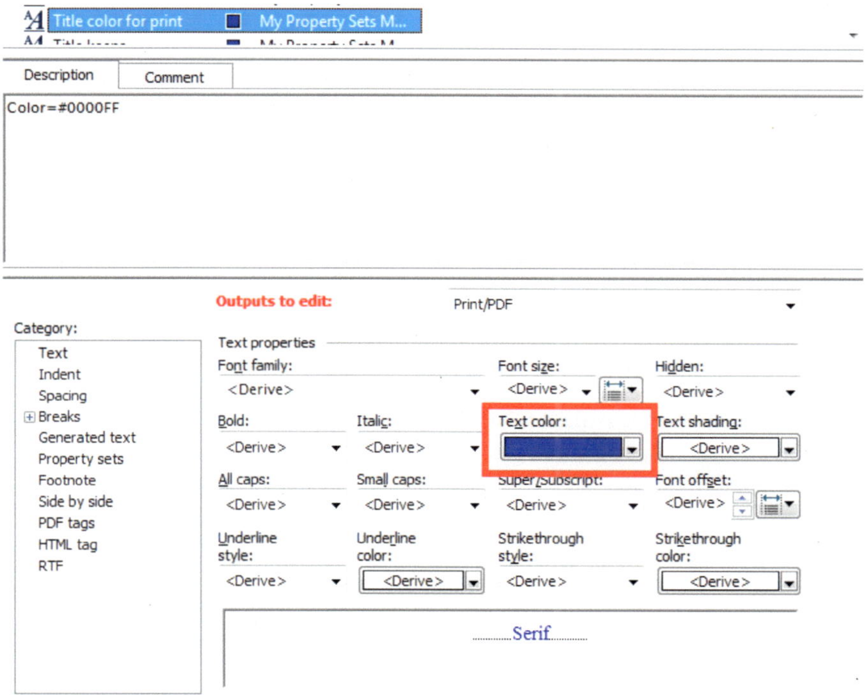

None of the properties in this panel are blue, either. It seems like a contradiction but it's not.

When you were looking at the context rule and changed the **Outputs to edit** to **Print/PDF**, that setting carried over when you switched to the property set rule. Even though we are looking at a property set, the **Outputs to edit** is still set to **Print/PDF**.

If you look at the **Description** pane, you can see that this property does not have output-specific formatting settings. No specific outputs are called out in the description panel.

Change the Outputs to edit back to Base (All Outputs).

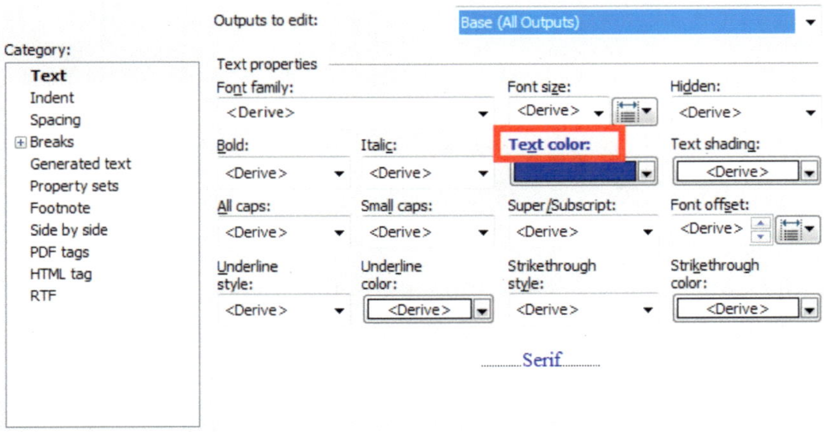

Voila! The **Text color** property has a specific setting: **Blue**. You just needed to make sure you had the right output format selected.

Copy the Title color for print property set rule to your My Property Sets Module.

Change it to **<Derive>**. You don't need to specify black because the default text color is black. When you change this property setting to **Derive**, it will use the default text color and make no changes.

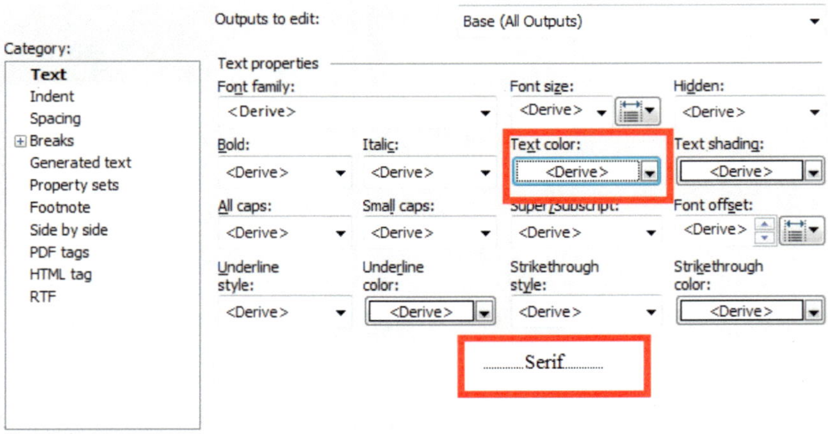

Notice that now there are no **blue** property names. And the text in preview pane is black.

That takes care of the Print/PDF output, but you need to check the other formats of the **title** element rule. Click on the **Elements** tab and locate the **title** rule. Change the **Outputs to edit** to **Base (All Outputs)**.

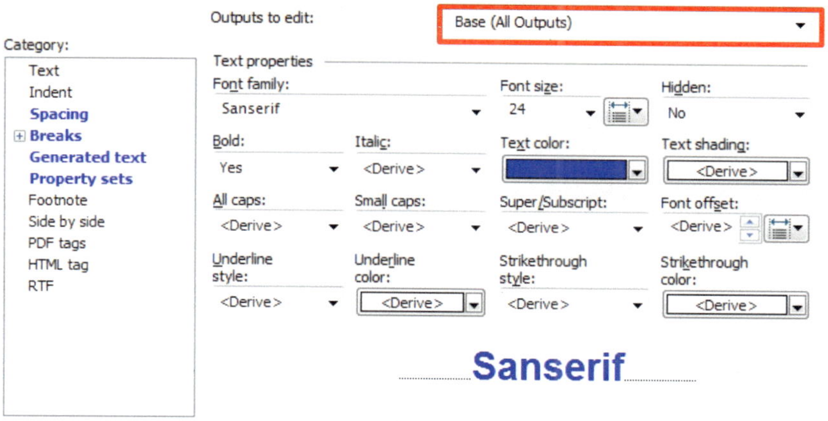

Notice that the color in the preview is still blue. Even though the **Text color** property name is not blue, the setting is for a blue color. Time to go hunting again. Start with the **Property sets** (they're blue).

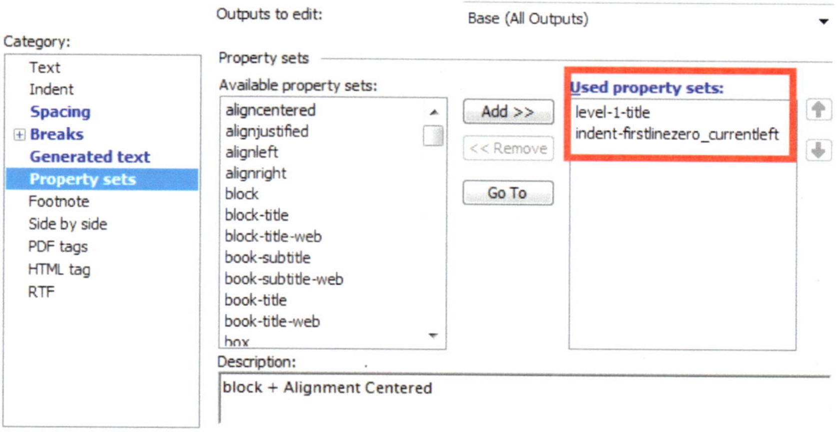

Two property sets are listed here:

- level-1-title
- indent-firstlinezero_currentleft

The property set with font settings is likely the first one, **level-1-title**. The second one has a name that suggests **Spacing** and indent settings rather than font. Select **level-1-title** and click **Go To**.

The **level-1-title** property set is comprised of more property sets.

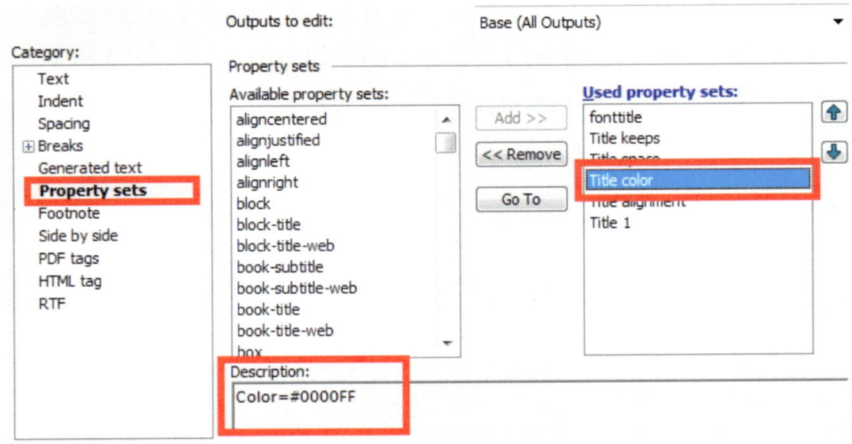

Click each one and check the **Description** field to learn about what each one does and what properties it affects.

When you click on **Title color** property, you can see in the description that it has a color setting. Click the **Go To** button.

Copy the **Title color** property set rule to your **My Property Sets Module**.

Change the text color to **default** just like you did for **Title color for print**.

Be sure to check the other output formats to make sure none of them others have a color setting. You shouldn't need to make any more changes, but once you've verified the other output formats, you can address numbering and labeling formatting.

Change the generated text

Go back to the element context rule for **title in chapter**.

The **Generated text** category is blue, so you know there are changes set on this rule. Before you can start changing it, remember that this element rule has output-specific formatting. You need to see if the generated text category has output-specific formatting. Change the **Outputs to edit** to **Print/PDF**.

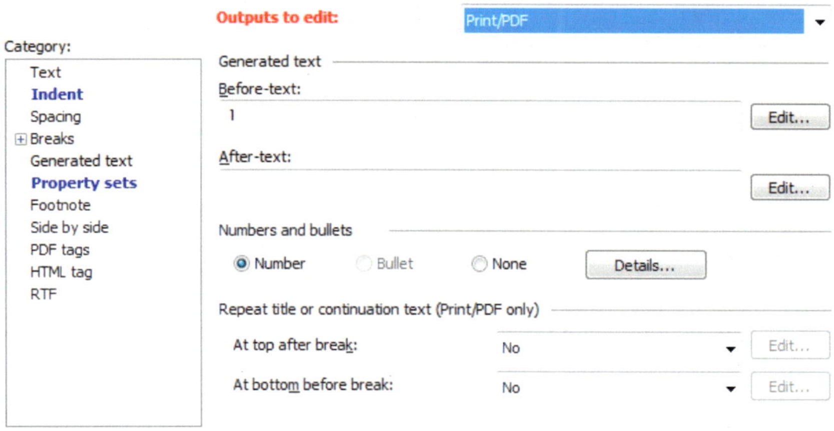

In the **Print/PDF** view, the **Generated text** category is not blue and none of the properties in the generated text category are blue either. That means you need to switch back to **Base (All Outputs)** to make your changes.

Change the **Outputs to edit** to **Base (All Outputs)** and select the **Generated text** category.

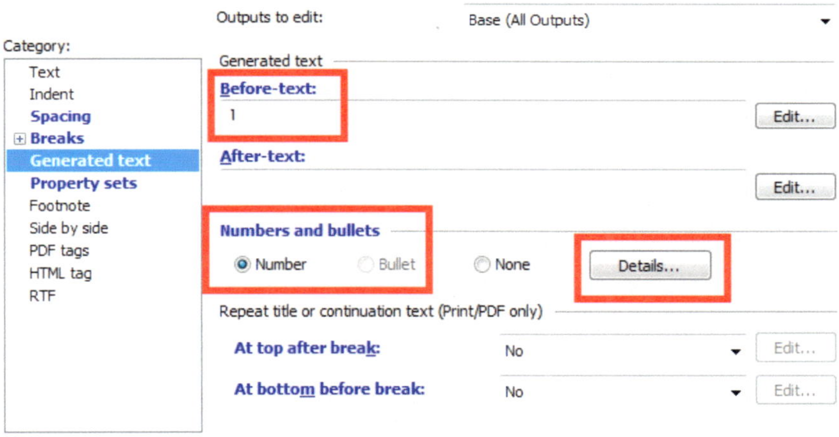

This category can be deceptive to novice and advanced users alike. In the **Before-text** property, you can see a preview of what the generated text looks like: **1**. You want it to say **Chapter 1**. While it might seem that you should make the change in the **Before-text**, you actually want to make the changes through the **Details** button, because this title has **Numbers and bullets** applied. The **Label** is part of the number generation and not typed before-text.

Click **Details** to open the **Division Title Number** window.

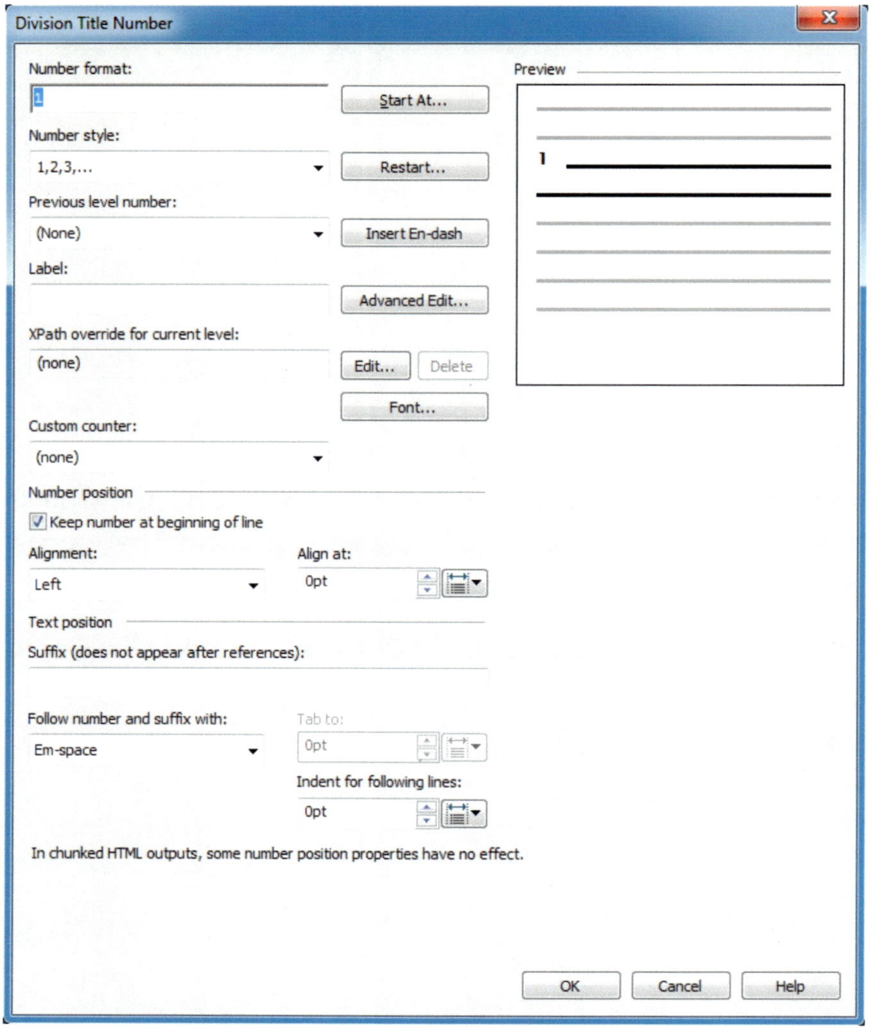

This window has a field for **Label**. Type **Chapter** and a space character in the **Label** field. You can see the effect in the **Preview** pane.

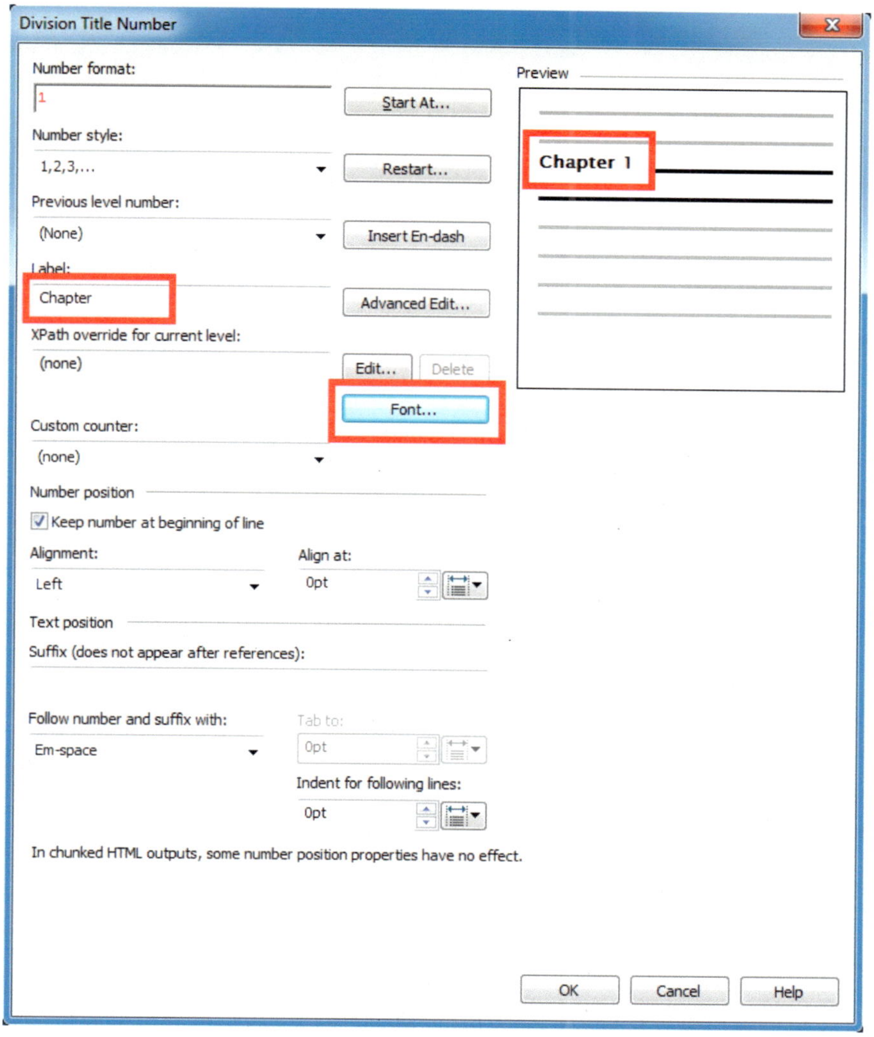

Click **Font** to change the font size to **Inherit**. You want the label and chapter number to be the same size as the rest of the chapter title.

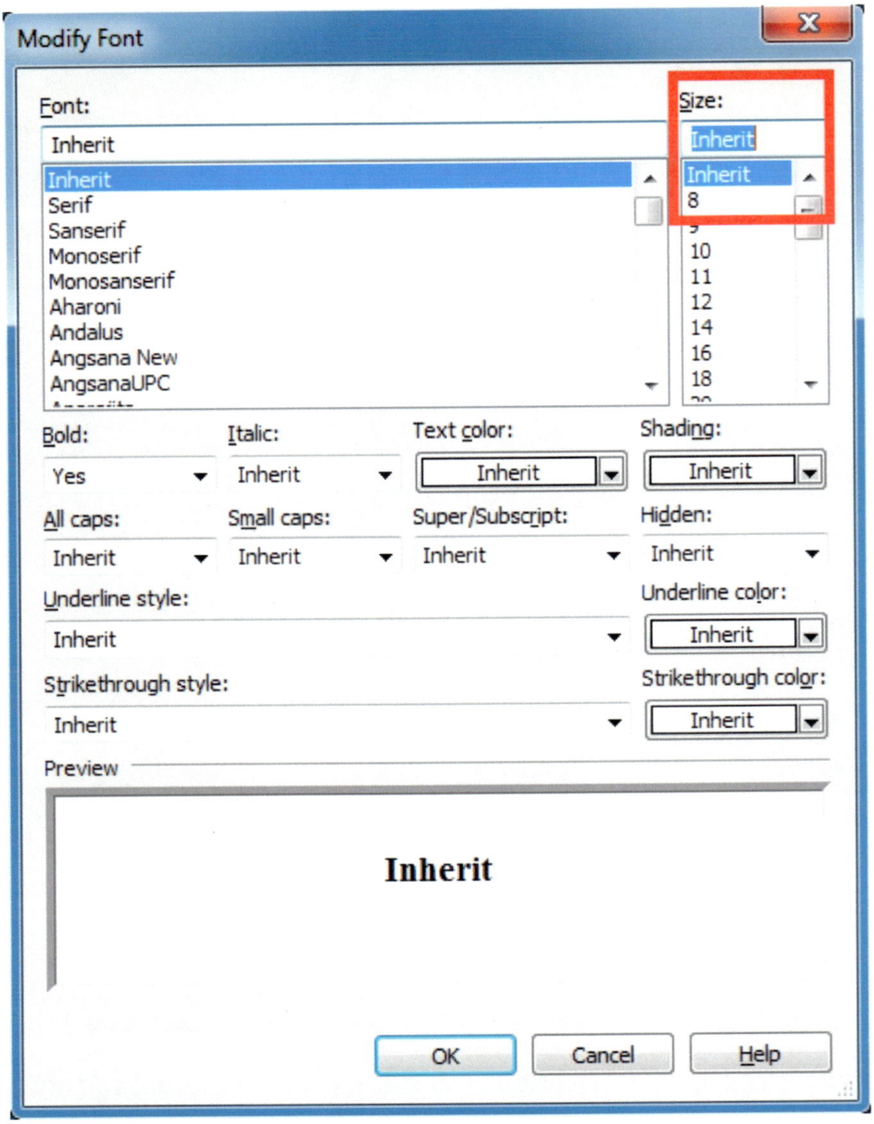

It's now time to test your changes. Click the **PDF** icon in the stylesheet editor toolbar. When the PDF is generated, you will see that it now says **Chapter 1** and the color is black (not blue). The size of the label and chapter number are the same as the rest of the title characters.

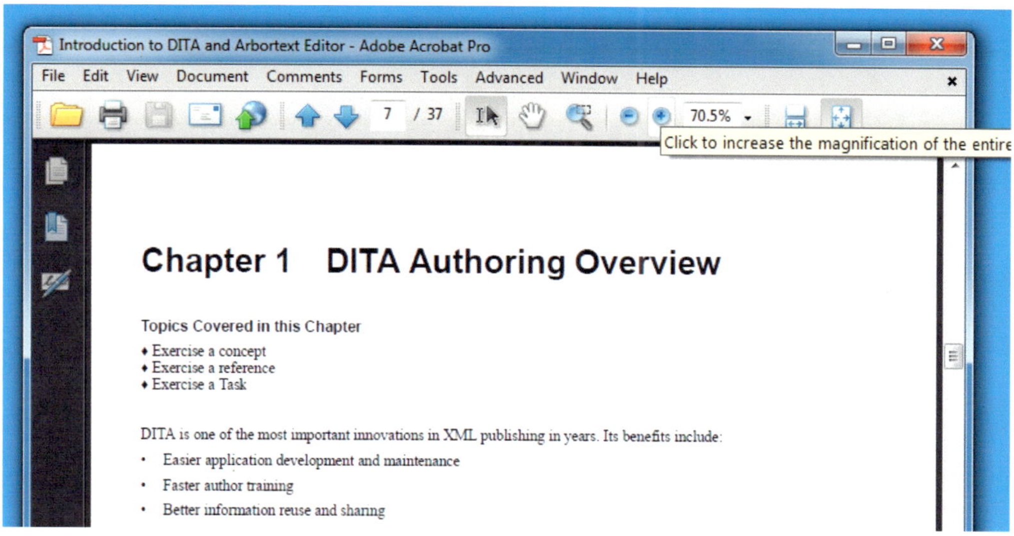

Save your stylesheet.

Expert tip!

Why did changes to the **Generated text** category in **Base (All Outputs)** affect the PDF output? Because changes made to **Base (All Outputs)** can apply to all outputs. Unless a setting in a specific output contradicts the base, the settings for the base apply to all output formats

This is another form of inheritance operating in the stylesheet. The base serves as the default formatting setting. If you want to make a change for a specific output format, you change the **Outputs to edit** and then make the change that should apply in that case or that should **override** the base setting. More specific trumps default. If there is not a more specific setting, the default property applies.

How to remove numbering from sections

Removing numbering from sections is similar to what you did for chapter. If you look at the PDF created from the sample document, you will see that sections that are children of chapter (heading 2) have numbers like the chapter had numbers. Sections that exist in appendix all have numbers, regardless of level.

Chapter Page

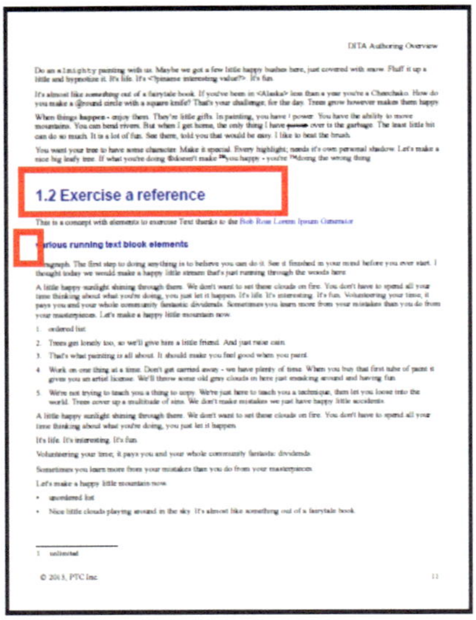

Appendix Page

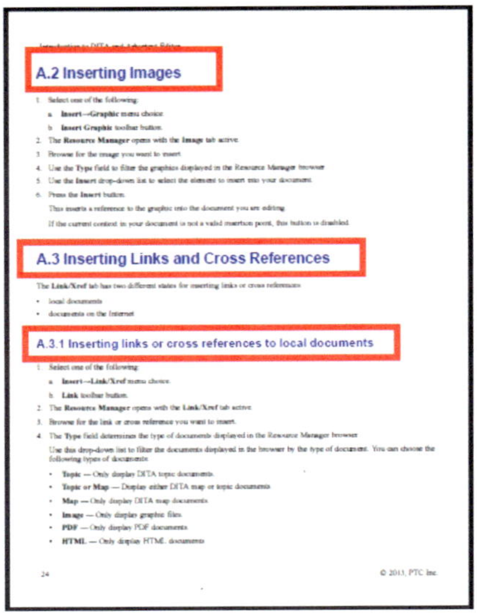

If you want numbering in the appendix, you can leave the appendix as is. You can also use the appendix-related context rules as the example to add support for numbering to all titles in a chapter. Or, you can remove inconsistent numbering from the heading 2s in chapter.

I'll walk you through how to remove the numbering from heading 2s in a chapter and leave the appendix for you to do on your own.

The first step is to figure out exactly what context rule represents heading 2.

Start at the sample document. In the document editor, try to figure out where **Exercise a reference** appears.

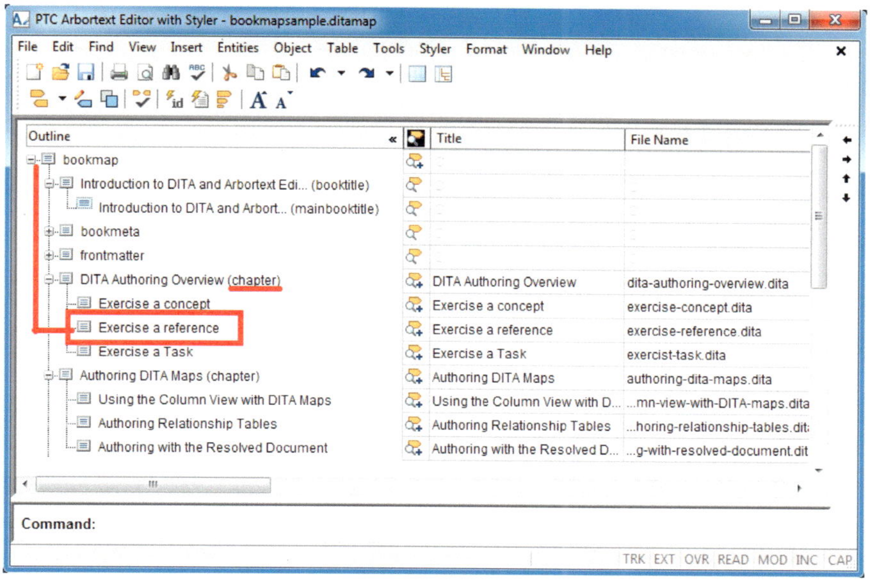

Exercise a reference is a title in a topic that is itself a child of a chapter.

The context you want is **title in topic in chapter**. Find that context rule in the stylesheet.

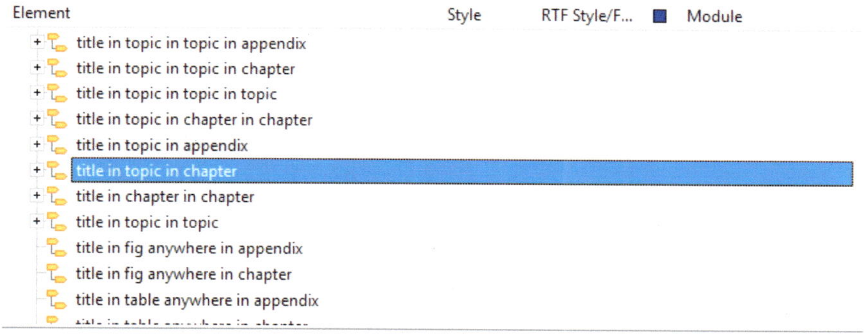

In the **Description** tab, you can see output-specific settings. You can also see that this context is considered a heading 2. All of the rules have some setting with **level-2-title** in the name.

Follow the same procedure you did for **title in chapter**. First, start with the **Print/PDF** output and see if you need to change any settings. Numbering is part of generated text, so you should look all the places where the **Generated text** category name is blue.

The outputs that have specific settings for **Generated text** are:

- EPUB
- HTML Help
- Base (All Outputs)

In both the EPUB and HTML Help outputs, the numbering and bullets are set to **None**.

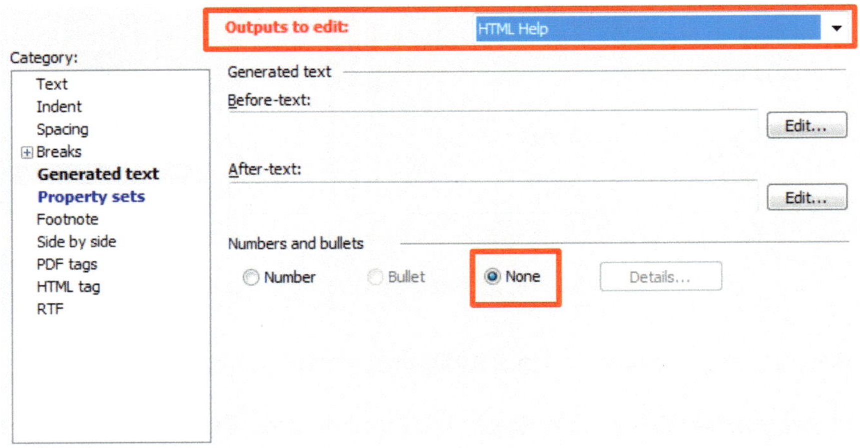

They are overriding the behavior of the Base. You know this because the Base has numbering properties set.

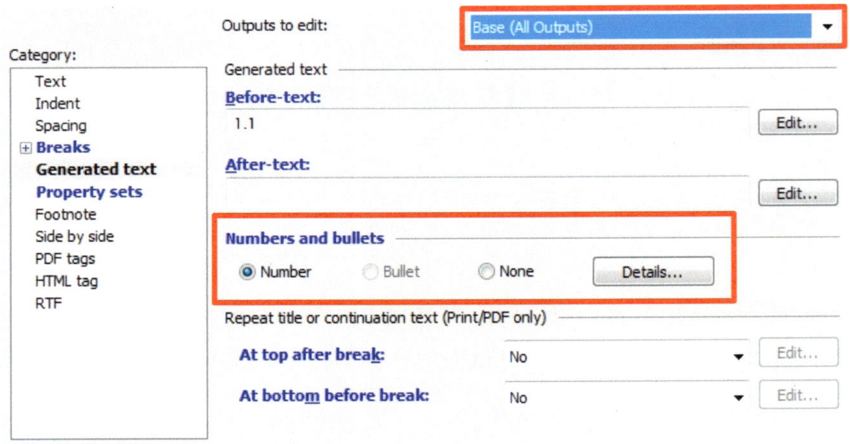

To remove the numbering, click the **None** option and Styler will automatically remove all the settings related to numbering.

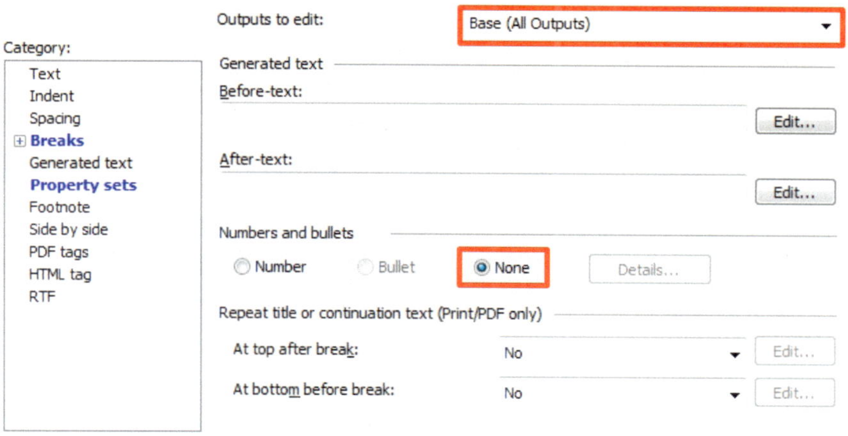

If you want to remove numbering from the titles in the appendix, you will need to find all the contexts that apply, such as:

- **title in topic in appendix** (heading 2)
- **title in topic in topic in appendix** (heading 3)
- **title in topic in topic in topic in appendix** (heading 4)
- **title in topic in topic in topic in topic in appendix** (heading 5)
- and so on…

You might also want to address **title in appendix** if you want to change the label so it says **Appendix A** rather than just **A**, and address the size of the **A** as well.

Save your stylesheet.

Expert tip!

If you want to change the font properties of your headings, you should investigate context rules that represent each heading level. Walk through the property sets until you find the setting you want to change. Change it in the property set so the change is uniform and universally applied to all headings that appear at that level.

How to fix the titles in the preface

The OOTB stylesheet does not include formatting for the preface. If you look through all 51 context rules for title, you will not find one for **title in preface** (a heading 1 equivalent to a **title in chapter** or **title in appendix**). You need to add a context rule for this condition.

In the stylesheet editor, select one of the context rules for the **title** element and click **Insert Context**. In the New Context window, click **New Parent** and type **preface** in the field.

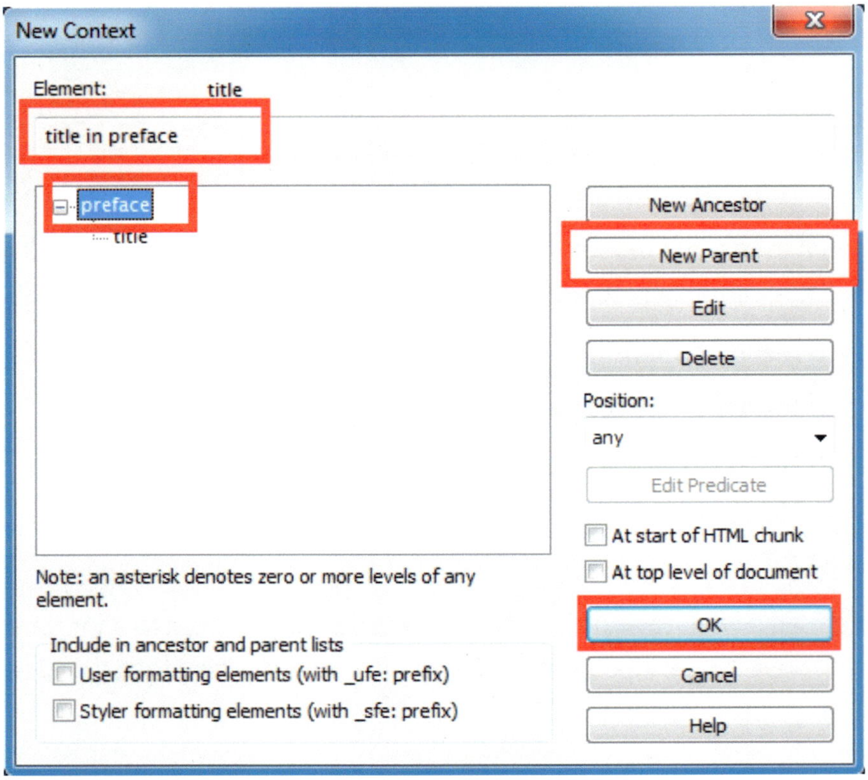

You will see that the description of the new context changes to **title in preface**. Click **OK** to close the window.

You might see a warning message:

Don't worry about this message. This is a valid context, as we know. Styler is validating against the ditabase and not the bookmap DTD. Preface is a bookmap element.

Your context rule will be added to the list. You already know what property sets to add. You saw them when you were working on **title in chapter**.

To add a property set, first locate the property set in the **Available property sets** list, then click **Add** to add it. You can remove property sets you don't want by selecting the property set in the **Used property sets** list and then clicking **Remove**.

Remove the property sets that Styler added by default when it created your new context and add these two in this order:

1. level-1-title
2. indent-firstlinezero_currentleft

If you want to add support for headings 2, 3, 4, and 5, be sure to add those context rules and make them look like the corresponding rules for those heading levels in chapter.

Save your stylesheet.

How to change to multi-column layout

You can configure a page set such that the same page set can be used for multiple-column layout. You can predefine what the page set looks like for 1-column layout, 2-column layout, and so on.

The element rule that triggers the page set can specify how many columns it wants to use with that page set.

The OOTB stylesheet creates the Index as two-column layout. The **indexlist** element rule is where you triggered the **Index Page Set**. Although you specified the page set, you never specified the number of columns.

Knowing now how inheritance works, it's time to reexamine the **indexlist** element rule.

Locate the **indexlist** element rule and select the **Breaks** category in the lower pane. Expand all the contexts and conditions and one-by-one look at the settings in the **Breaks** category for each context and condition.

Wherever you see it set to **Columns** 2, change it to **<Derive>**.

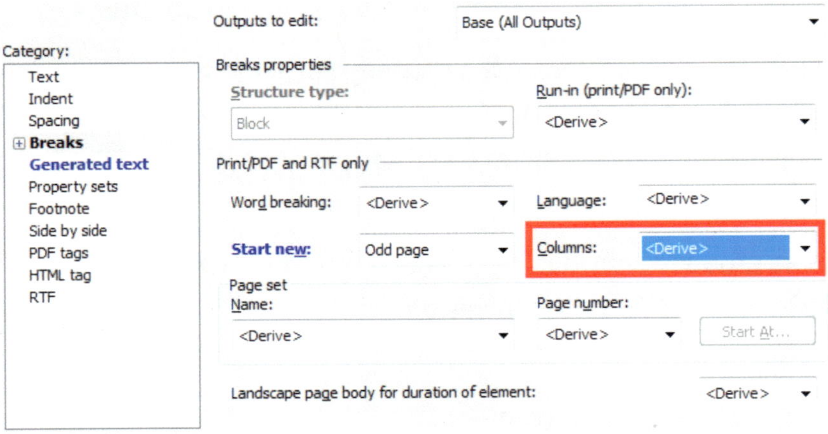

Be sure to check all the contexts for each specific output format. Some of the contexts and conditions of the **indexlist** element rule have output-specific settings. If you generate a PDF and still see two columns after you think you have set everything to **<Derive>**, I can guarantee you missed a setting that applies only to a specific output format, a context, or a condition.

Save your stylesheet.

Expert tip!

It is always better to set a property to **<Derive>** to than to give it a specified value. When a rule derives the property setting, you are *unsetting* that property (clearing the specification). By unsetting, you won't have to track down a property to fix later. You want to be able to depend on settings, which is easier when you can count on them being derived.

How to set base document properties

Now that you know how the different kinds of inheritance work and how element rule processing works, you should already have some idea about how and where to set basic formatting properties.

The more specific and the more localized a specific formatting property is defined, the higher priority it has. A child element inherits the properties set on its parent, but the child element can override that inherited value.

Remember how this worked with chapter title. The format for the label was different (bigger font) from the rest of the title format. The label belongs to the title; it is processed at the time the title is processed. The formatting applied to the label had a higher priority because it was local to the label. When you removed the local formatting on the label, it inherited the same formatting as the rest of the title content. Local formatting has priority over inherited formatting: The label format overrode the chapter title format.

Inheritance can make it a little harder to track down where a property is set so you can override it correctly, but it can also work in your favor.

If you have a font family and standard size for running text that you want to apply uniformly throughout your document, you can set it once—on the parent-most element rule—and have every element inherit it. You can then override that setting where needed: you can make apply local formatting on a chapter title overriding the base font settings set for the entire book.

For any formatting properties that you want to serve as your defaults, you set them on the parent-most element for your documents. In the case of the sample files, that's **bookmap**.

Locate the **bookmap** element in the stylesheet editor and take a closer look at how formatting is applied to this element.

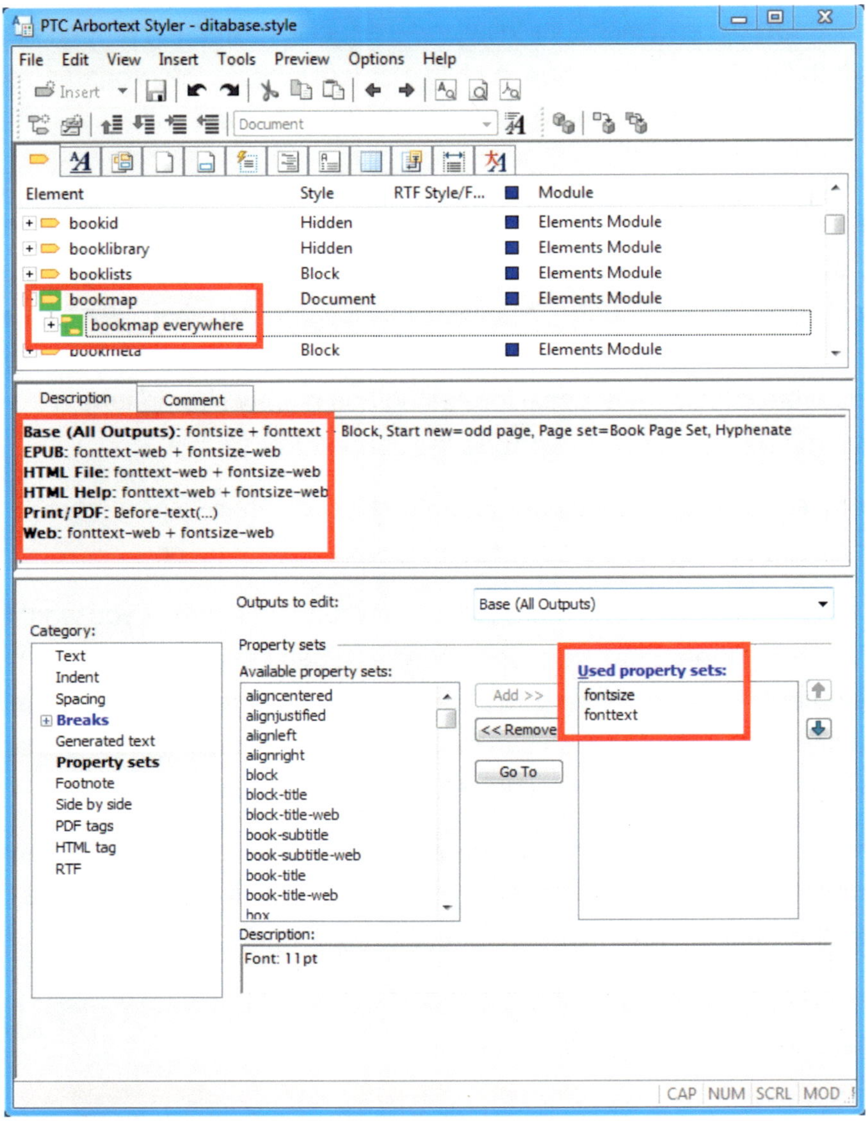

The **bookmap** rule has two property sets assigned: **fontsize** and **fonttext**. This rule also has contexts and conditions, as well as output-specific formatting. You will need to check each one and make sure that there are no specific settings for font size or font family on any of the rules in each output-specific format.

You also want to look closely to see if other property sets appear in any of the contexts, conditions, or output-formats. Make a list of all the property sets that are used anywhere in this rule. Then examine the property sets themselves.

You can change the base font and base size simply by changing the property sets.

> **Expert tip!**
>
> Before you make a change to any rule don't forget to copy it to your module first. You don't want change the OOTB modules. Save your changes and your work. Always make changes to rules that are safely stored in *your own modules*.

User Formatting Elements (UFEs)

User Formatting Elements (UFEs) are elements that you create in your stylesheet and insert into generated text for the purpose of organizing and controlling the application of formatting properties. A UFE can enclose any generated text, including elements and element content. You can create as many UFEs as you want to support the formatting requirements of your stylesheet. UFEs give you the ability to consolidate formatting properties.

Like property sets, a UFE gives you the ability to encapsulate formatting into a single element that can then be inserted around generated text, giving you a way to isolate a group of format settings that you can deploy in multiple places. When you change the UFE, it changes the formatting for all of the generated text enclosed everywhere the UFE is used.

For most stylesheet designers, UFE are an unusual concept. I'll use a couple of cases to demonstrate the power of using UFEs. First, I'll readdress the headers and footers we created earlier: I had you create them, but I didn't to show you how to format the header and footer generated text. UFEs can help you simplify formatting tasks in your own stylesheet.

UFEs are grouped under the **Elements** tab. They are given a **_ufe:** prefix automatically by Arbortext Styler to distinguish them from regular element rules. In addition, UFE element rules, context rules, and conditions have blue tag icons instead of yellow ones.

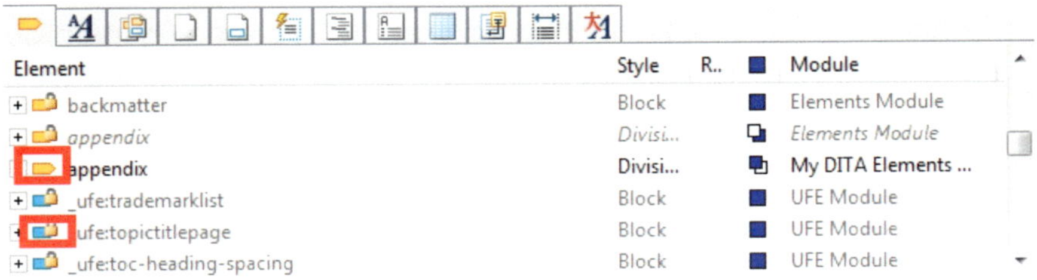

Formatting generated text in headers and footers

Headers and footers are a part of a page, but are processed separately from elements in the body of the page. Font settings that you set by default on your parent-most element (**bookmap**) do not apply to page set regions.

The best way to apply standard formatting properties to the content in your header and footer is to enclose the text of the header and footer in a UFE.

Click the **Generated text** tab, select the **Chapter Right Page Header Content** rule, and click **Edit** to open the edit window.

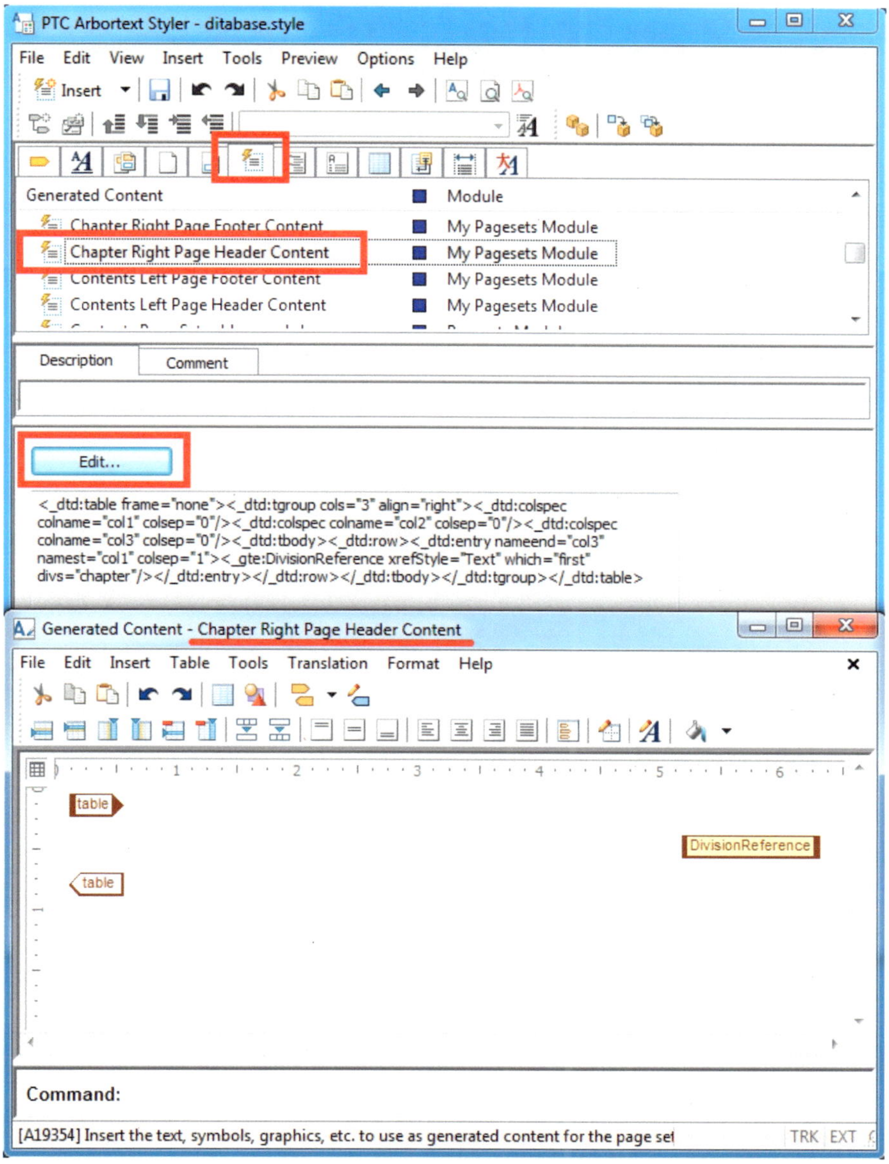

Right now, the header has a table, and will display the content of the most recent division reference. No formatting information is applied in this window. There is no element rule that matches **DivisionReference** because it isn't an element in your source document. The Header resides strictly within the stylesheet. To format the header content, you need to insert a UFE around the text in this window.

You can either insert the UFE around the **table** or the **DivisionReference**. Either way works equally well.

If you look at the **Chapter Right Page Footer Content** rule, you have multiple text elements in two different columns of the table.

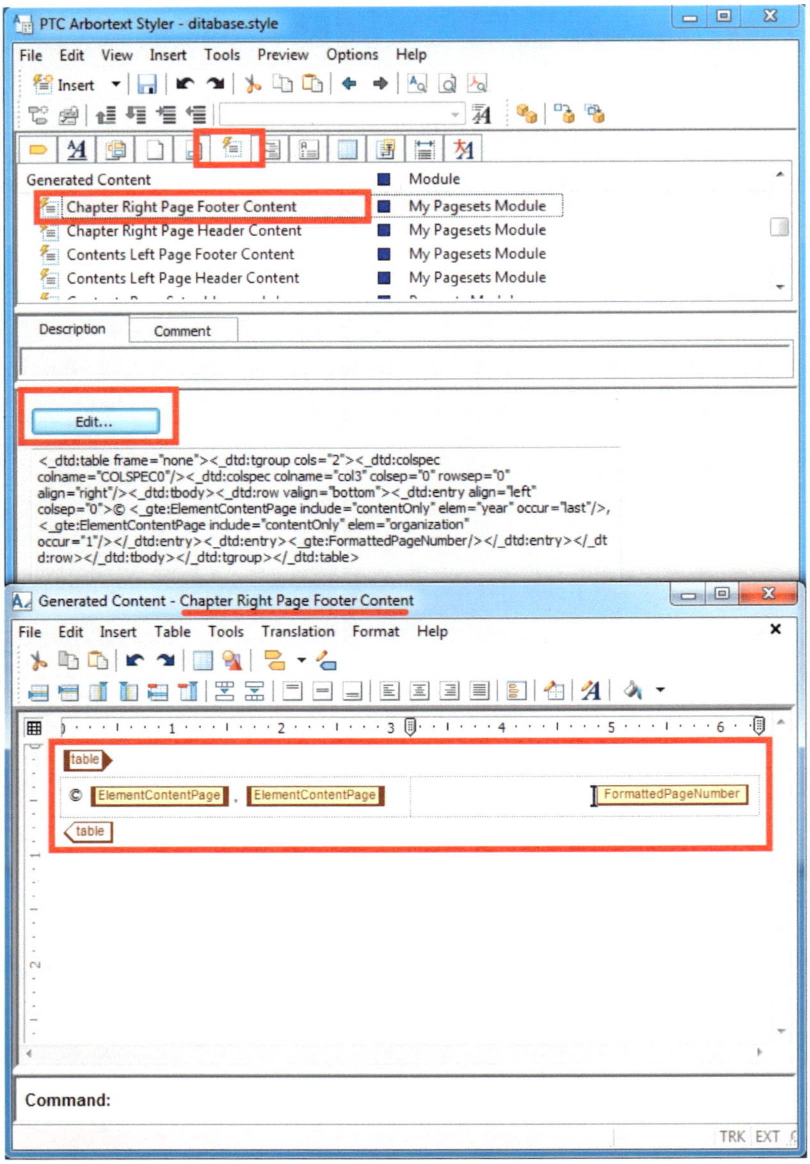

In this case, it's easier to wrap the entire table. Your UFE becomes the highest-level element for all the generated text that appears in the table. We recommend being consistent in your application. Wrap the individual text elements or wrap the larger context (**table**).

Return to the **Chapter Right Page Header Content** rule and open the edit window. Select the table.

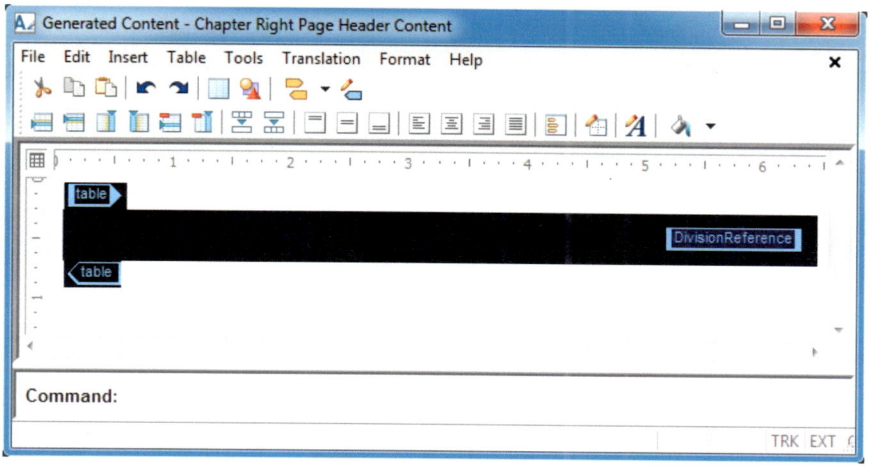

Select **Insert→User Formatting Element→(new)**.

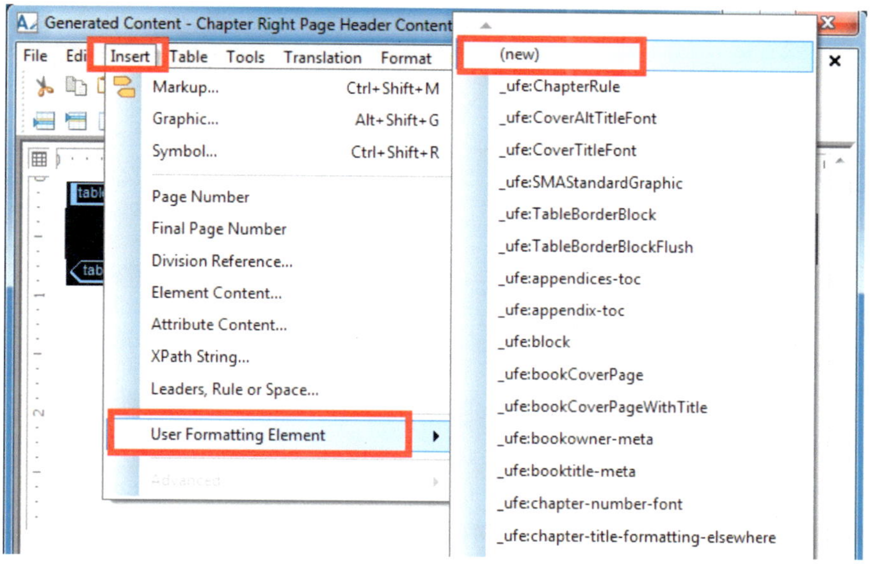

In the **New User Formatting Element** window, give your new UFE a name: **HF_format**.

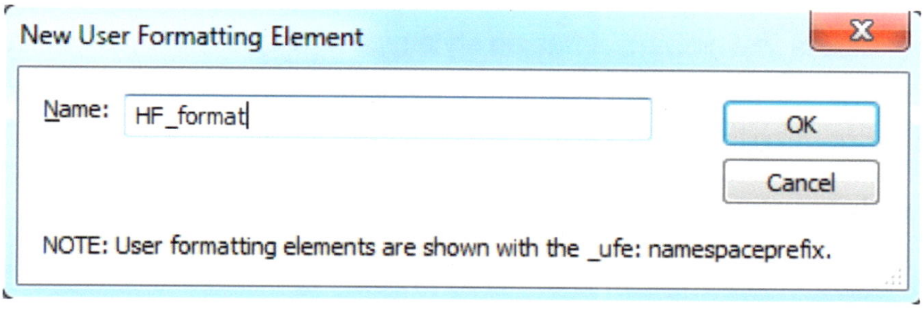

Click **OK** to return to the edit window. Arbortext Styler will automatically insert this new element around the **table** element, enclosing it and all of its contents.

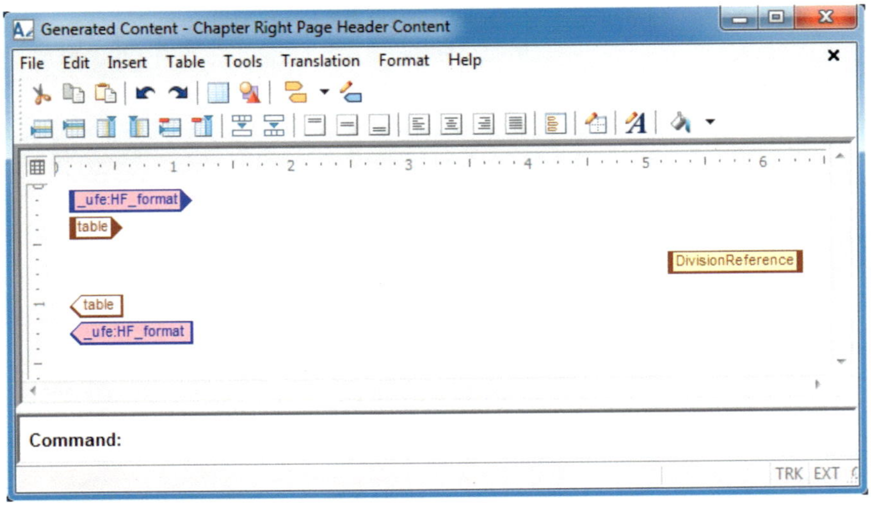

Go to the **Elements** tab and locate the UFE you just created:

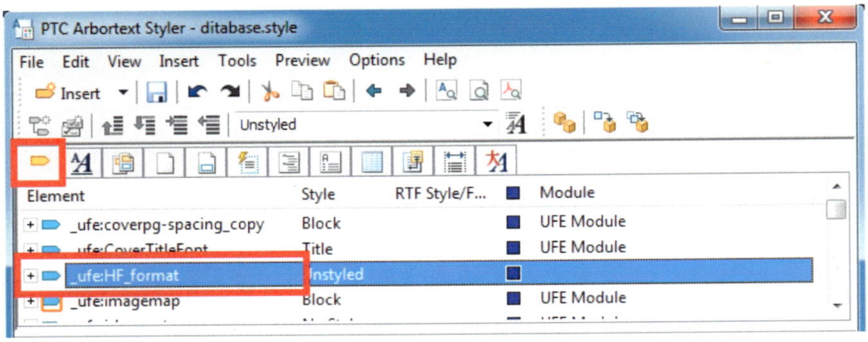

You have actually created an element. The content of the header and footer resides inside an element that gets processed by the rendering engine and the stylesheet. Once you give your UFE a style type, you can assign formatting properties to it, like any other element in your stylesheet or source document.

When a UFE is first created, it has no style type because it is a new element that you created. Arbortext Styler needs you to assign a style type so it knows how to categorize this element.

Right-click on the rule and choose **Style→Block**.

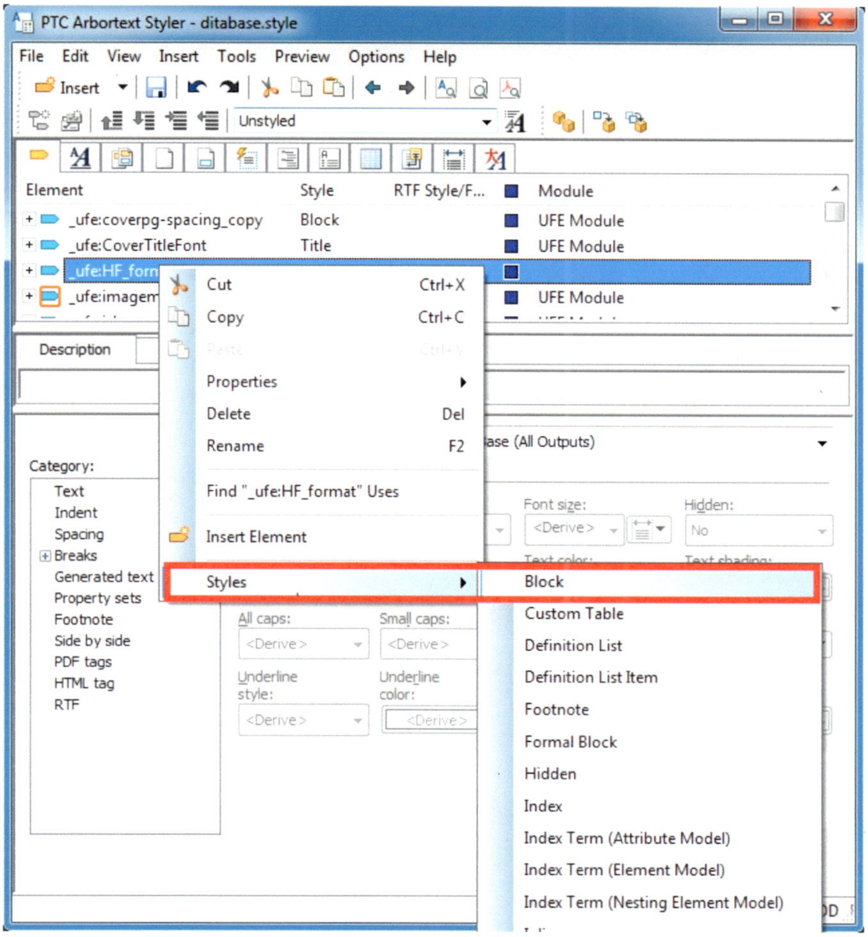

The header is a block on the page, so the **Block** type is the most logical choice. Now add the two property sets that represent the standard font family and font size—**fontsize** and **fonttext**—to the property sets list for this element rule.

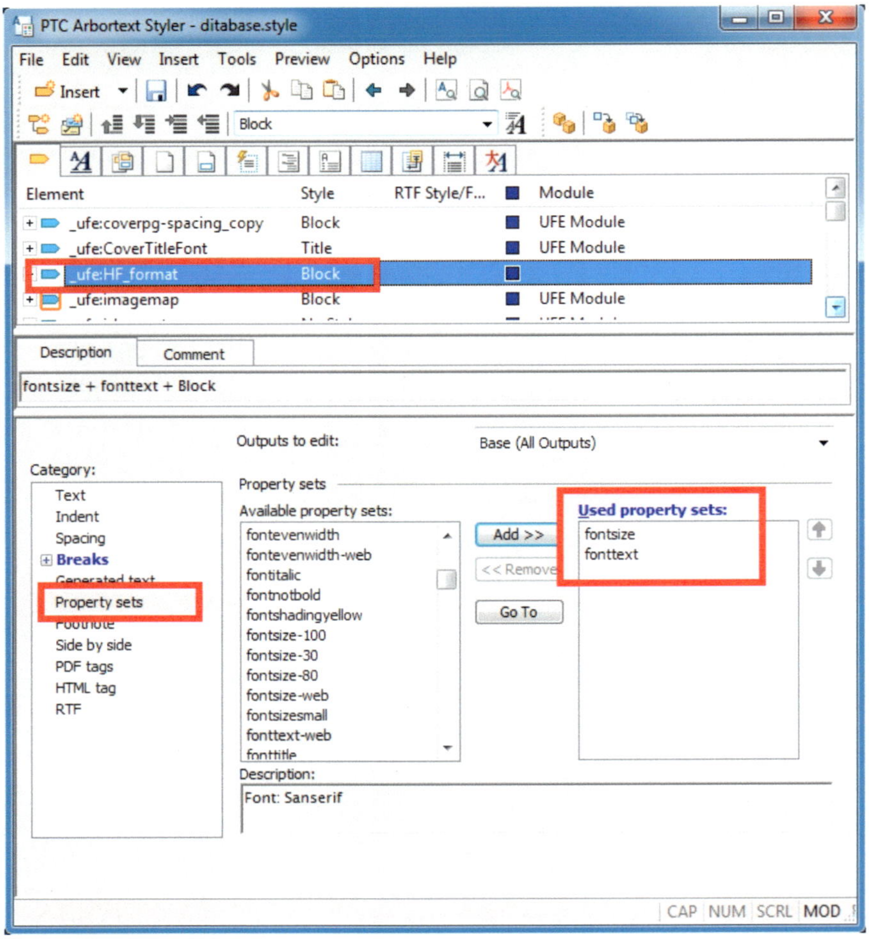

If you want your header to be bold, add **fontbold** to the list. Add any property sets you want to this UFE.

Go the **Chapter Right Page Footer Content** rule. This time, when you go to the **Insert→User Formatting Element** menu, your UFE will appear in the list of UFEs you can insert. Choose your UFE from the list.

Repeat the same steps for all the other header and footer content rules.

UFEs give you a lot of power. With a UFE you have one place to encapsulate formatting properties that you can then use to apply formatting automatically wherever you use it. You don't have to format every text instance by hand. And when you want to change the format of your headers and footers later, you change the properties on the UFE. It will change the formatting everywhere it's used.

Expert tip!

Use UFEs wherever you want to localize and encapsulate formatting in an organized way. But UFEs are most powerful when you see them as elements that can be processed like any other element in the document. When you create a UFE, you create an element out of nothing. You can create a UFE that inserts content from elsewhere in the document. You can create a UFE and then use it in another element's **Before-text** or **After-text** to insert and format content from elsewhere.

UFEs are strictly stylesheet inventions that give you the flexibility to create elements. With UFEs, you can create formatting situations that you can manipulate as if they were simply another element in your source document.

UFEs are the key to advanced formatting techniques that stylesheet designers employ regularly. A UFE is the stylesheet designer's secret weapon to manufacture structures and content as needed.

UFEs as a context for doctype elements

Locate the element rule for **mainbooktitle**.

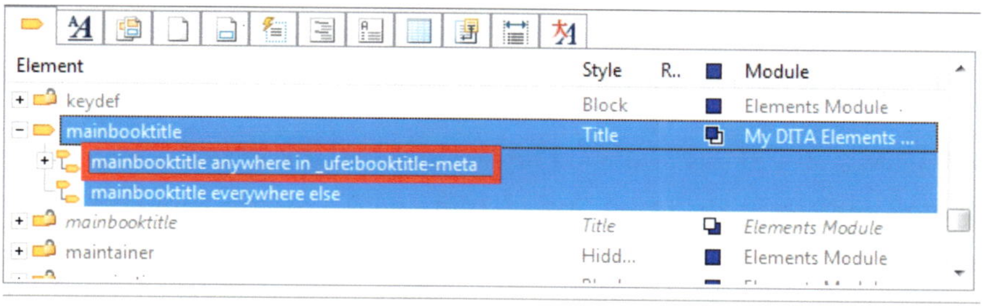

One of its contexts is **mainbooktitle anywhere in _ufe:booktitle-meta**. This is possible because of the way generated content works.

Remember how you constructed the header. To display the book title in the left-hand header, you inserted the content of the **mainbooktitle** element in the header. But that's not the only option. You could have inserted the element as well as its content.

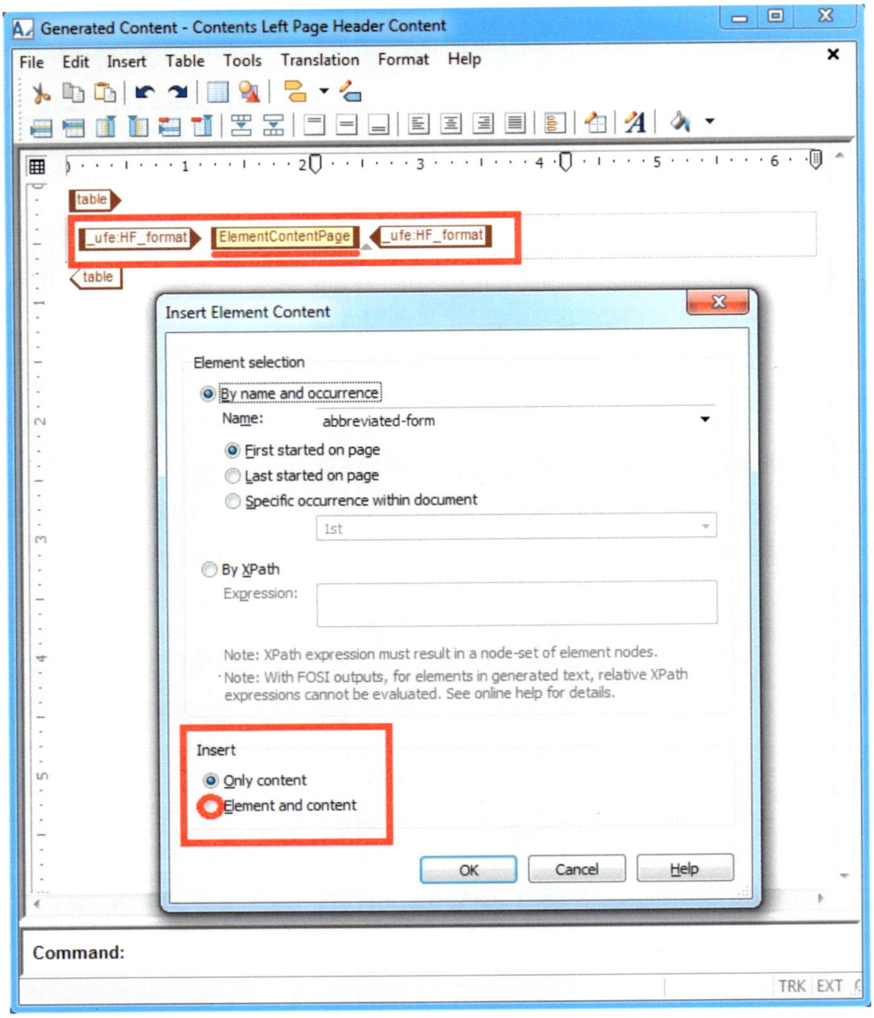

If you had inserted the element and its content into the header, you created a new context when you wrapped that element with the UFE that you created, **HF_format**. You can see the markup in the generated text editor. Because you created this new context, you can add it to the element rule for **mainbooktitle**. You can treat the context as if it were any other context and apply formatting to it, including output-specific formatting. It's a powerful option used by advanced stylesheet developers.

Chapter 9. Creating Covers

Topics Covered in this Chapter

♦ Preparing to design the cover
♦ Prepare to create the cover page
♦ Investigate the elements that show on the cover
♦ Add formatting to cover elements
♦ Add content to the back of cover page

Once you understand the basics of page layout—how regions work, region content, generated content, and triggers, and property sets and UFEs—covers are a snap!

Here are the front and back of the OOTB cover:

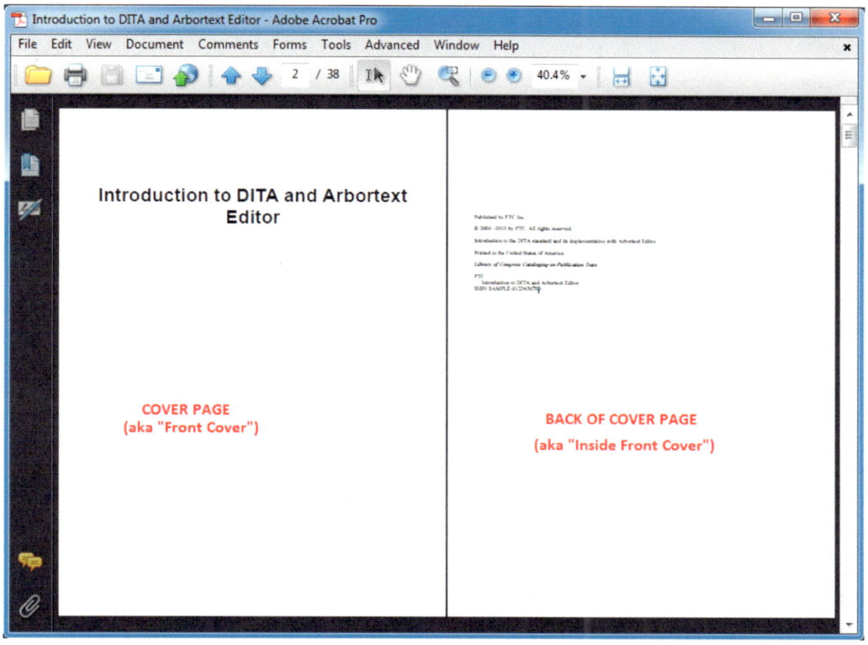

The title is on the front cover. The copyright and ownership metadata appears on the back of the cover page. The content on the back of the page does not start at the top of the page—it is actually somewhat lower than the top of the title line on the front cover.

Here's the cover you will to create:

The cover has two regions—header and footer. The header and footer on this page are much bigger than the header and footer on a standard page. The cover has a logo region that overlaps the header, putting the logo at the top of the page. The book title is located in the center, between the regions. The title is indented on both the left and right. There is a 6pt black line on the border of the graphic in the header and footer. This line separates the graphic from the center area where the book title is displayed.

Preparing to design the cover

Before you can work on a cover page you need to understand which elements trigger a new page, what page set is active, and how you want to construct the content on the pages.

Start with the source content:

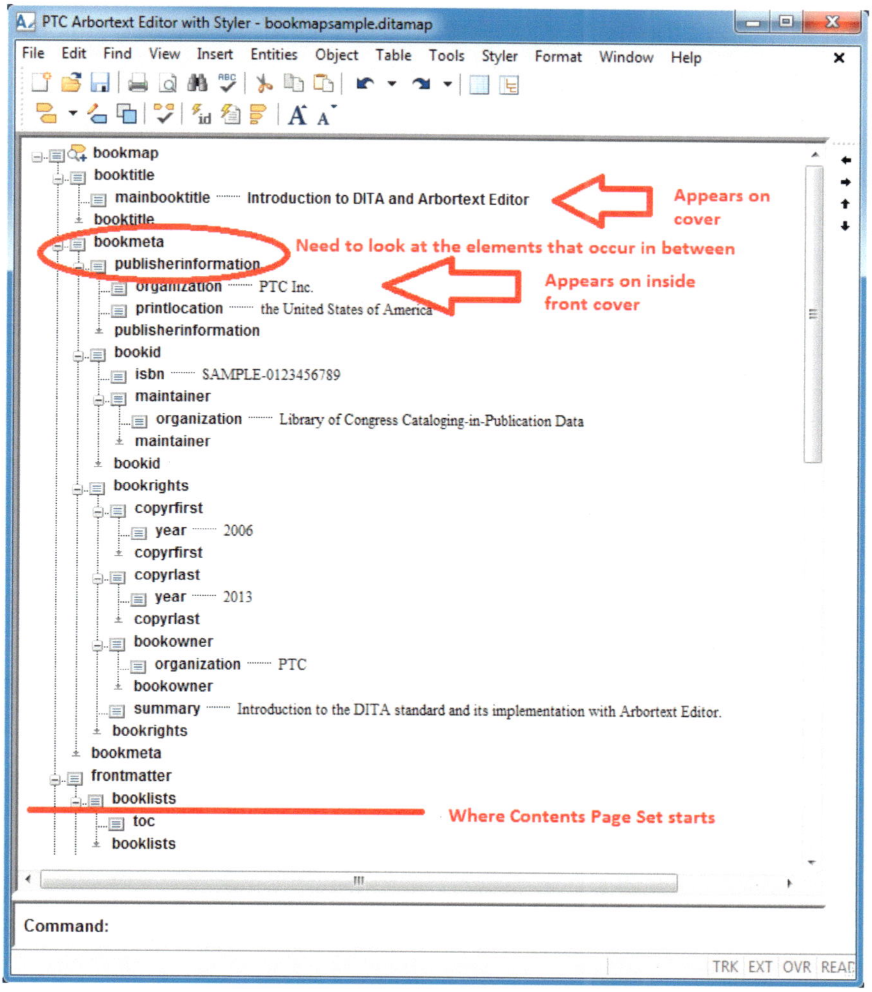

Here's what you know:

- The **toc** element starts the **Contents Page Set** and a new page
- The content in **mainbooktitle** appears on the cover
- The content in **organization** appears on the back of the cover.

The content that can appear on the back of the cover must appear before the **toc** element, which indicates the start of the **Contents Page Set**.

There are two elements between the content that you know appears on the front of the cover and the content that you know appears on the back of the cover: **bookmeta** and **publisherinformation**. Either of these items could be the one that triggers the new page, causing content to appear on the back of the cover instead of the front. Examine both element rules to see which element is responsible for starting a new page.

The **bookmeta** rule has special formatting applied to the **Print/PDF** output format: It triggers a **new page**.

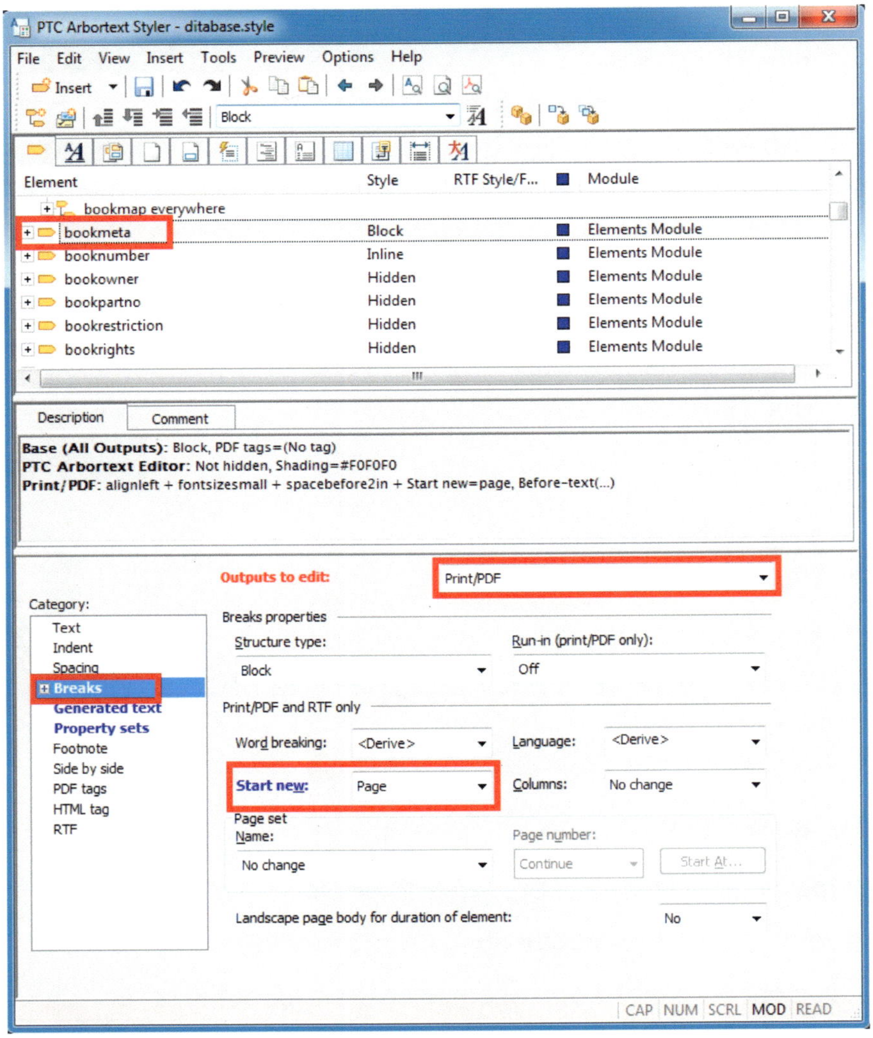

Now that you know where the dividing line is between the front and back of the cover page, you should examine each of elements before and after that dividing line.

The following elements that are used to construct the front and back of the cover page. They determine the order in which their content appears on the page:

1. bookmap
2. booktitle
3. mainbooktitle
4. bookmeta
5. publisherinformation
6. organization
7. printlocation
8. bookid
9. isbn
10. maintainer
11. organization
12. bookrights
13. copyrfirst
14. year in copyrfirst
15. copyrlast
16. year in copyrlast
17. bookowner
18. organization
19. summary

Examine each element rule to see which rules have the following items:

- Output-specific formatting
- Contexts
- Conditions

- Property sets
- Generated text (Before-text and After-text)

For each property set, follow it through to see how many property sets include other property sets and discover what each one does at each step.

For the generated **Before-text** and **After-text**, see if UFEs are used, where they are used, and investigate each one to discover what it does. Examine each UFE rule the same way you would any element rule: UFEs are just another type of element rule.

It is a lot to investigate. Remember that the OOTB stylesheet is as much an example of what is possible as it is a foundation for you to build upon. In the case of the cover, that is doubly true. For now, look at the different element rules and property sets to get a feel for how everything works. I'll walk you through setting up the cover page, but you will need this investigative process in the future when you're coding your specific requirements later on.

Prepare to create the cover page

Using the skills you've learned so far, create the cover page. Remember to move any new pages and regions you create to your **My Pagesets Module**.

Create three new page regions as follows:

Region Name	Settings
Cover Bottom	**Category**: Position • **Region reference point**: Bottom Left • **X position**: From page left, 0in • **Y position**: From page bottom, 0in • **Width**: Page width, +0in • **Height**: Absolute, 3.5in **Category**: Graphic • **Underlay region text with graphic**: Check this box • **Graphic path**: yellow.jpg • **Horizontal scaling**: Fit region • **Vertical scaling**: Fit region **Category**: Borders • **Top**: Check this box • **Thickness**: 6pt **Category**: Other • **Background color**: Green • **Underlaid regions avoid this region**: Check this box
Cover Top	**Category**: Position • **Region reference point**: Top Left • **X position**: From page left, 0in • **Y position**: From page top, 0in • **Width**: Page width, +0in • **Height**: Absolute, 3.5in **Category**: Graphic

Region Name	Settings
	• **Underlay region text with graphic**: check this box • **Graphic path**: yellow.jpg • **Horizontal scaling**: Fit region • **Vertical scaling**: Fit region **Category**: Borders • **Check**: Bottom • **Thickness**: 6pt **Category**: Other • **Background color**: Green • **Underlaid regions avoid this region**: Check this box
Cover Logo	**Category**: Position • **Region reference point**: Top left • **X position**: From left margin, 0in • **Y position**: From page top, 0.75in • **Width**: Width between margins, +0in • **Height**: Absolute, 2in **Category**: Graphic • **Underlay region text with graphic**: Check this box • **Graphic path**: logo-ex.png • **Horizontal scaling**: Fit region • **Vertical scaling**: Preserve aspect **Category**: Other

Region Name	Settings
	• **Underlaid regions avoid this region**: Check this box

Create a new page type named **Cover page**. Add the following regions to it in this order (front to back):

1. Cover Logo
2. Cover Bottom
3. Cover Top
4. main content region

Go to the **Book Page Set** and change the **First Page when right** to **Cover Page**.

If you publish to PDF right now, you'll see the framework, and the regions appear as expected. However, the title won't appear in the center of the page. You configured the page framework, but not the element rule to lay content on the page. Right now, with no additional formatting on the **mainbooktitle** element rule, it starts right below the **Cover Top** page region because of two settings:

1. **Cover Top** region is "on top" of the main content region.
2. **Cover Top** was set such that the regions that are below it avoid laying content wherever the regions overlap.

The **main content region** is used on every page set you have. You want to avoid changing its size and position. It's better to tell regions to avoid each other and to use other formatting properties to get the desired behavior than to change the behavior of the **main content region**.

At the moment, the content of the **mainbooktitle** starts below the bottom edge of the **Cover Top** region. Once the **main content region** is no longer underneath, the rendering engine is free to start laying text on the page.

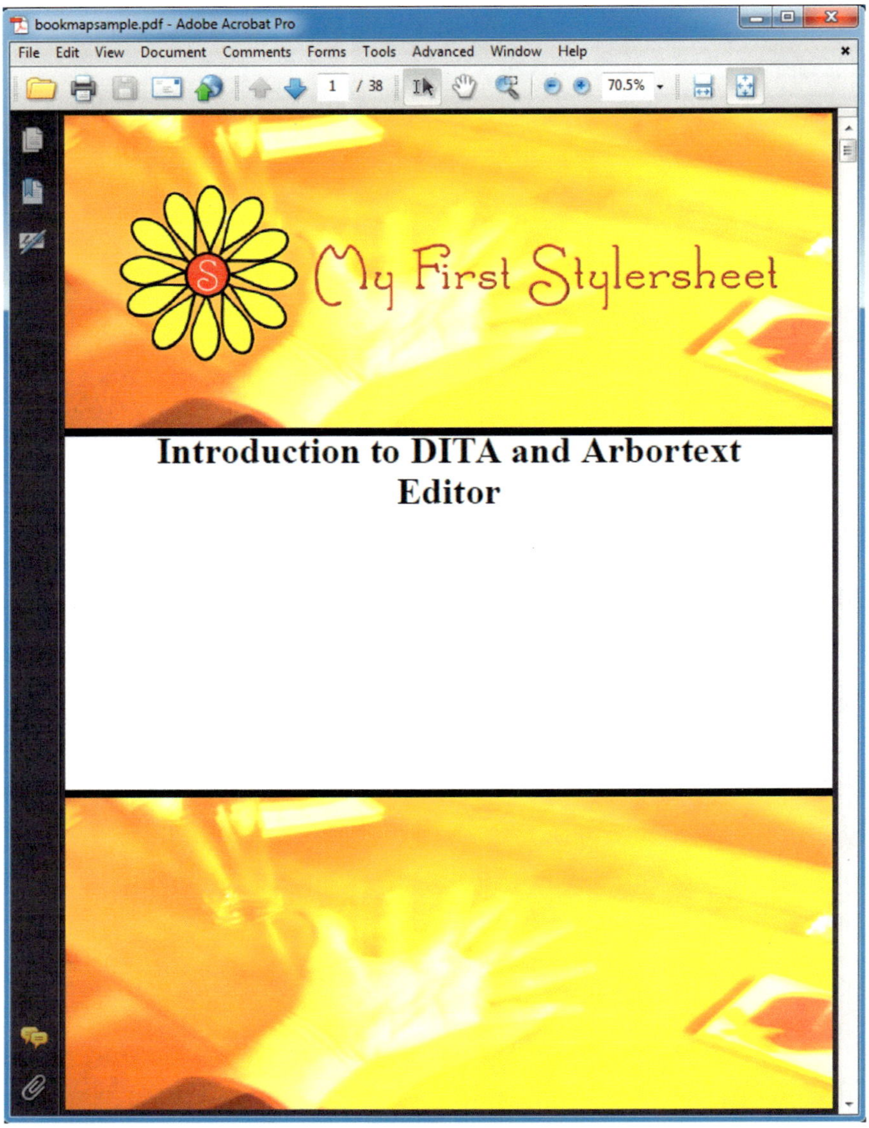

To move the book title farther down the page, you add *space above*. To make the title break differently, increase the left and right indents. You need three new property sets to control the space above along with the left and right indents.

Create these three new property sets in your **My Property sets Module** as follows:

Name	**Settings**
cover title space before	**Category**: Spacing • **Spacing before**: – **Preferred**: 1.5in – **Precedence**: high – **Minimum**: 1in – **Maximum**: 3in – **Keep space at top of column or page**: Keep
indent-left-1.5in	**Category**: Indent • **Indentation**: – **Left**: 1.5in **Relative to**: Left Margin
indent-right-1.5in	**Category**: Indent • **Indentation**: – **Right**: 1.5in **Relative to**: Right Margin

Expert tip!

The spacing category enables you the ability to set a preferred, minimum, and maximum space above and below. When you give the rendering engine a range of values, you increase its ability to help you in other ways, such as keeps (such as keep with next, keep with previous, or keep together) and widow and orphan control. When the rendering engine encounters a situation where it has to decide about breaking a page in a less than perfect manner, it will try to either squeeze or stretch the spacing to accommodate the best possible outcome. The values you give it for preferred, minimum and maximum spacing is what gives it that ability.

With automated publishing, you can't make manual page breaking decisions (and you don't really want to). If you are reusing content, a topic might break in one place in one document and in a different place in another. It's a good idea to give the rendering engine as much flexibility as you can.

Investigate the elements that show on the cover

You know that the mainbooktitle appears on the front of the cover page. In your source document, it occurs before the page break instruction tied to the bookmeta element. But one element occurs before it: bookmap.

Copy the **bookmap** element rule to your **My DITA Elements** module, and then expand its contexts and conditions. If you examine each context and condition individually, you will see that one context and one condition have **Print/PDF** output specific formatting attached to the **Generated text** category.

Change the **Outputs to edit** dropdown menu to **Print/PDF** and look at the context and condition again.

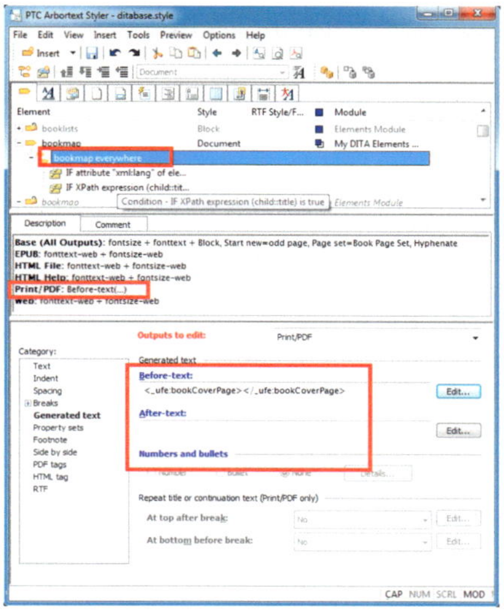

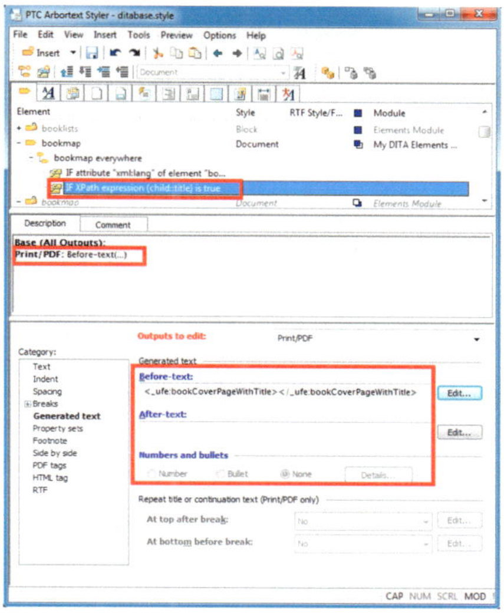

Each rule has a different UFE in the **Before-text** property. Instead of investigating the purpose behind using these two different UFE's here, clear the property setting.

Select one rule, right click the **Before-text** property value, and choose **Derive all generated text properties.**

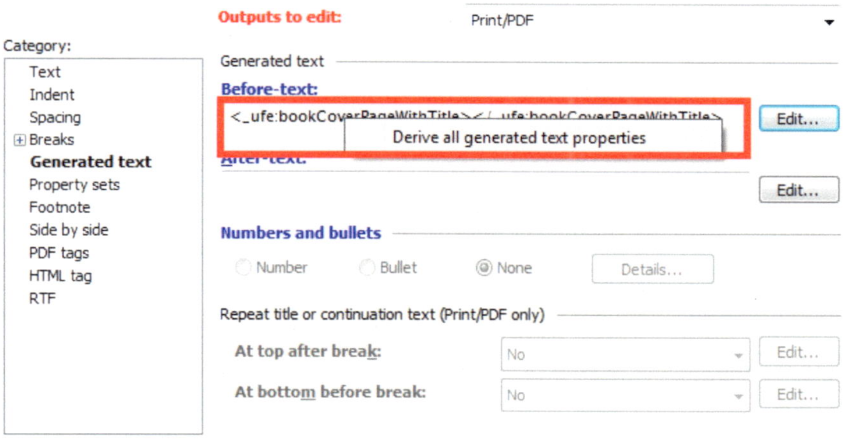

Repeat this step for the other rule.

When you tell a property to derive its formatting, all local settings will be cleared. Because this is the only category that had **Print/PDF** output-specific settings, the output-specific setting will also be cleared when you set the **Generated text** category to derive all properties.

You will have a clean starting point, and will avoid having to track down unexpected formatting that you didn't configure yourself. With a clean starting point, you can now examine the **mainbooktitle** element rule.

Locate it and copy it to your **My DITA Elements**. If you expand its contexts, you will see that this rule has two contexts:

- **mainbooktitle anywhere in _ufe:booktitle-meta**
- **mainbooktitle everywhere else**

If you click on the base rule, you see a couple of odd items: Orange italics on several property names and nothing in the Description panel.

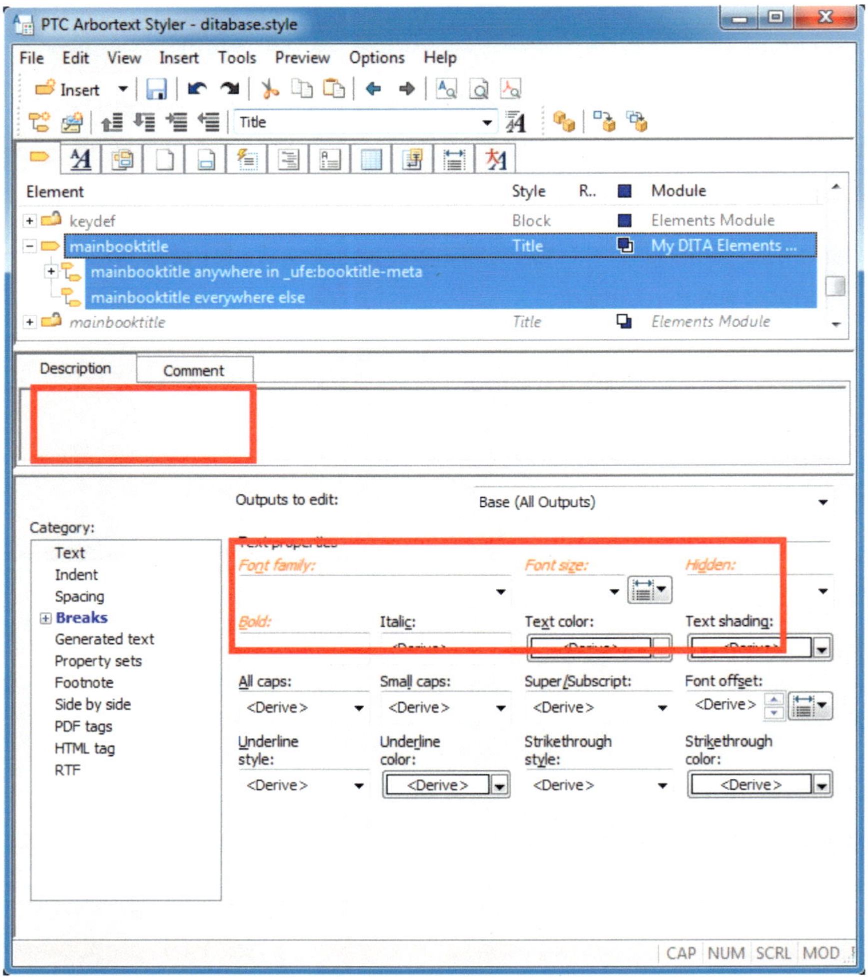

Orange italics mean that the settings for those properties are different in the two contexts. Changes to any of the orange-italicized properties in the parent rule will not push down to the contexts. If you want to make changes, you must change the individual contexts.

The first context—**mainbooktitle anywhere in _ufe:booktitle-meta**—is a stylesheet-only context because UFEs reside only in the stylesheet. Before taking any action, look at the UFE and find out where it's used. Locate the element rule for **_ufe:booktitle-meta** and select it. Choose **Find Where Used…** from the **Edit** menu to open the **Find Where Used** window.

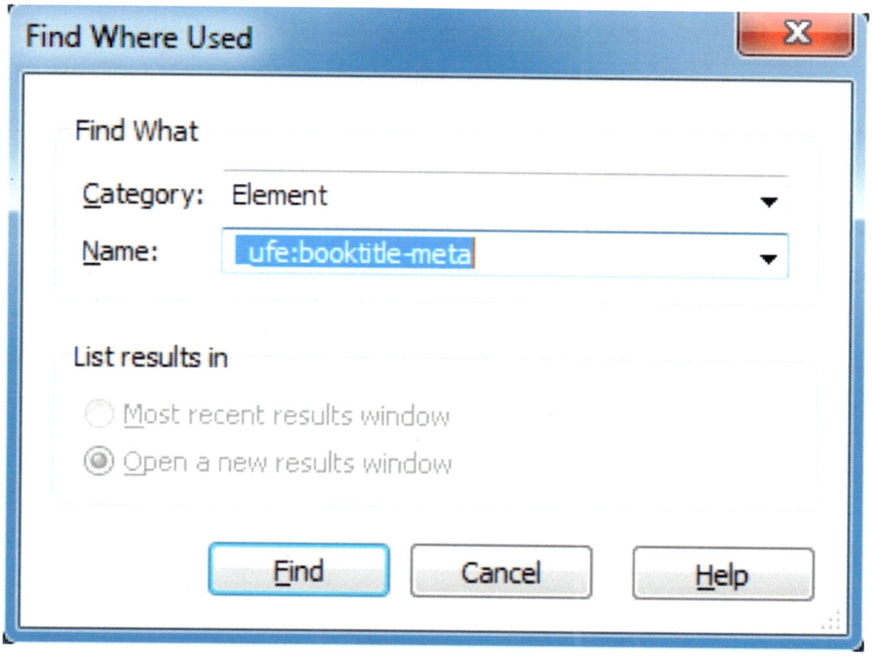

This window is pre-populated with whatever rule that was selected at the time. In this case, you selected an Element named **_ufe:booktitle-meta**. Click **Find** to open the results window.

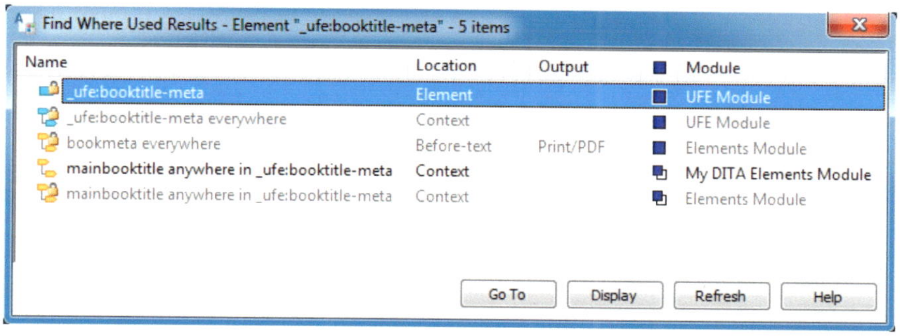

The results window displays a list of all the places in the stylesheet where this element is used, including rules in all the modules in your stylesheet.

The first two entries are the element context rules for this UFE. You know that because of the blue tag icons. The next three entries are all element rules. You can use this report to examine each rule. Select a rule to examine, and click **Go To**.

Of the three rules, one rule is overridden: **mainbooktitle anywhere in _ufe: booktitle-meta**. The active rule is the one in **My DITA Elements Module**. You already know about this instance. You went to investigate this UFE because you saw it in that element rule. This rule has priority over the duplicate overridden rule, so you can ignore it.

The **bookmeta everywhere** rule is already on our radar: This element starts a new page, the back of cover page. Right now it only resides in the Elements Module. You don't have a copy of it in your module, which means that it's currently active. And apparently it does a lot more than start a new page. Locate the **bookmeta** rule to examine it further.

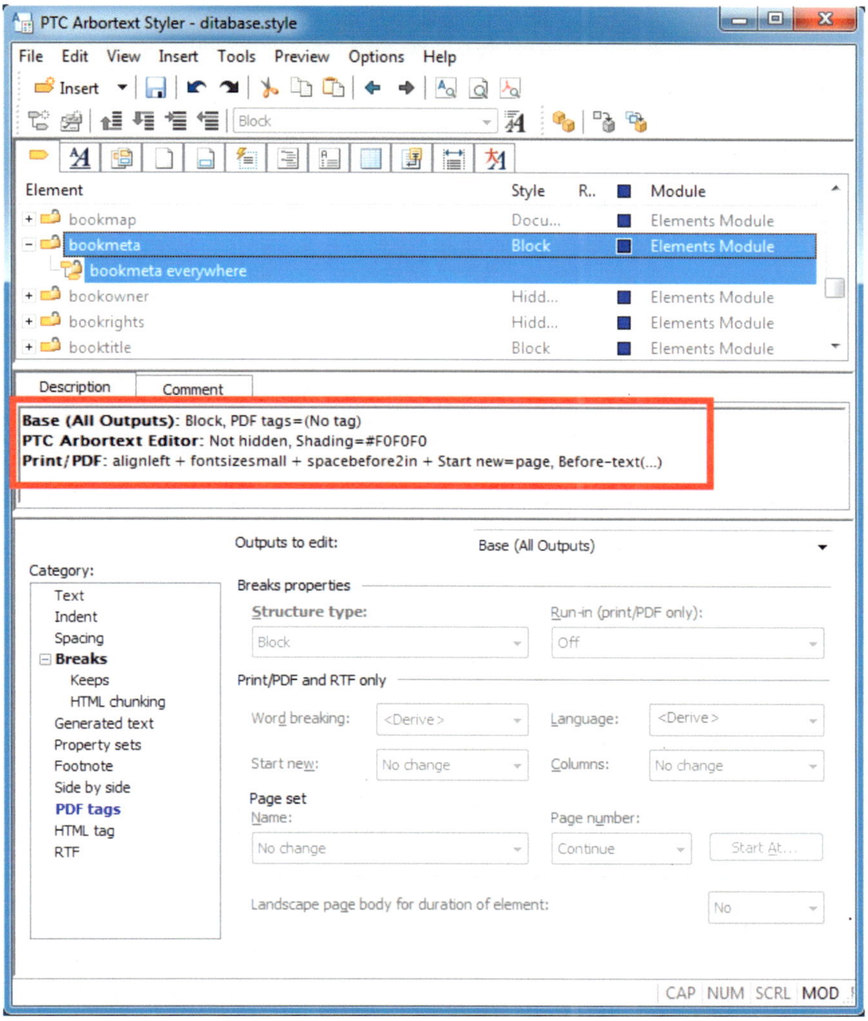

A complex rule to be sure because it has output-specific formatting applied. Change **Outputs to edit** to **Print/PDF** and select the **bookmeta everywhere** context rule.

Two categories are blue: **Generated text** and **Property sets**. By now, you should expect the **Property sets** category to be blue, so ignore it for now. The **Generated text** category is likely to yield more information about how the back of the cover page functions, since that's where we are likely to find the UFE in use.

Select on the **Generated text** category.

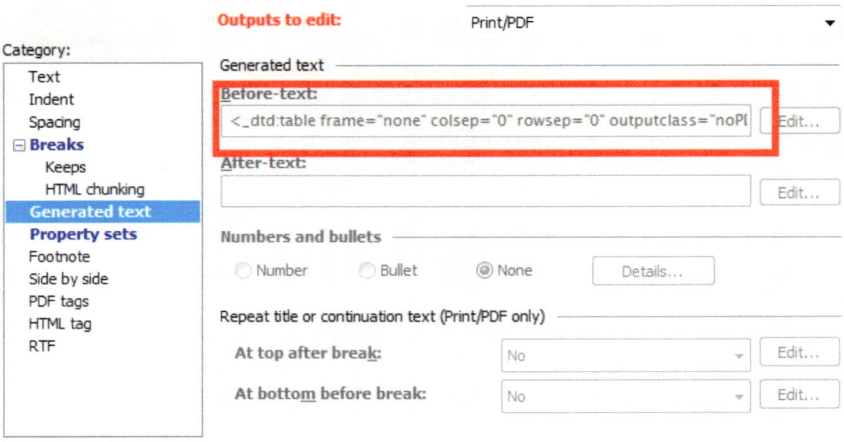

The **Before-text** property has some code. To get a look at it and make changes, copy this rule to your **My DITA Elements Module** and open up the **Before-text** editor. The editor has a lot of complicated markup. To find out exactly what's happening, you need to examine each of the **ElementContent** elements.

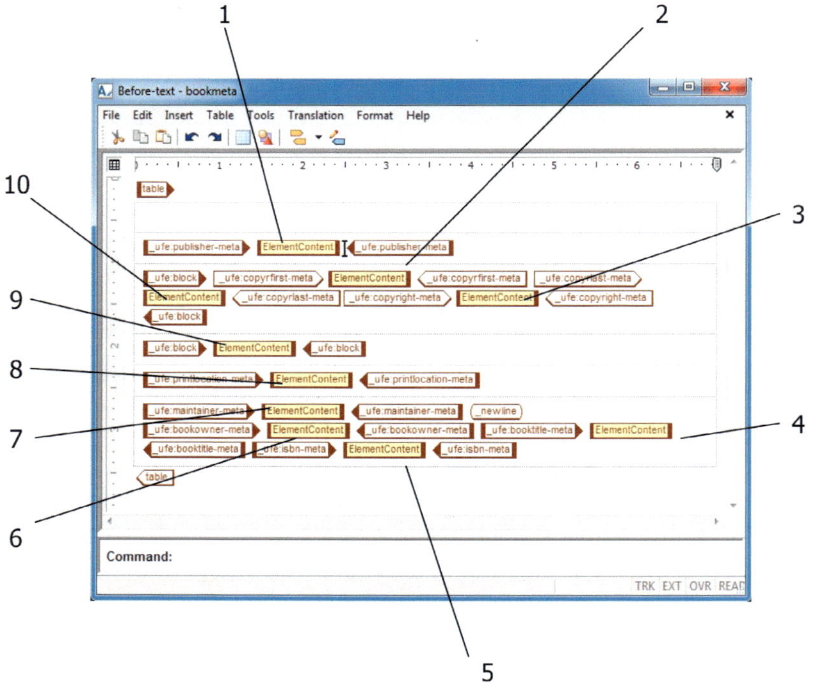

#	**What element content is displayed**
1	organization in publisherinformation
2	year in copyrfirst in bookrights
3	organization in bookowner in bookrights
4	first mainbooktitle in the entire document
5	isbn in bookid
6	organization in bookowner in bookrights
7	organization in maintainer in bookid
8	printlocation in publisherinformation
9	first summary in bookmeta
10	year in copyrlast in bookrights

Here, the stylesheet is generating the content on the page instead of processing the **bookmeta** element in place. The generated text has a table that contains content relocated from various elements into a specific order and position. Instead of processing the elements as they come, the stylesheet designer has chosen to create this structure and pull information from different elements, reorder them, and insert them manually into the table. This is called the *pull* method.

While a valid method, it's generally better to avoid using it when unnecessary. First, it's a lot of work. Second, it's hard to track down, as you just saw. It took quite a lot of investigating to find out that this page was being created in the first place, not to mention where and how it is being done. If you cannot depend on markup flowing in order of display, this context rule is a great example of how to do that.

However, it is a method to avoid whenever possible. Most of the time, you want content to be displayed in the same order it appears in the document. The first paragraph should be first. When the stylesheet handles each element in document order, formatting as it goes, this is called the *push* method. Even if you use the *pull* method for some parts of the book, most of your processing will be through the *push* method, so it's important to understand how to use it effectively.

I'm going to have you switch the back of cover page to the push method. Instead of constructing this page, you will process the **bookmeta** element and all of its children in order, formatting elements that you want to display and suppressing the elements you don't.

Close the **Before-text** editing window. Right-click the **Before-text** property field and choose **Derive all generated text properties**.

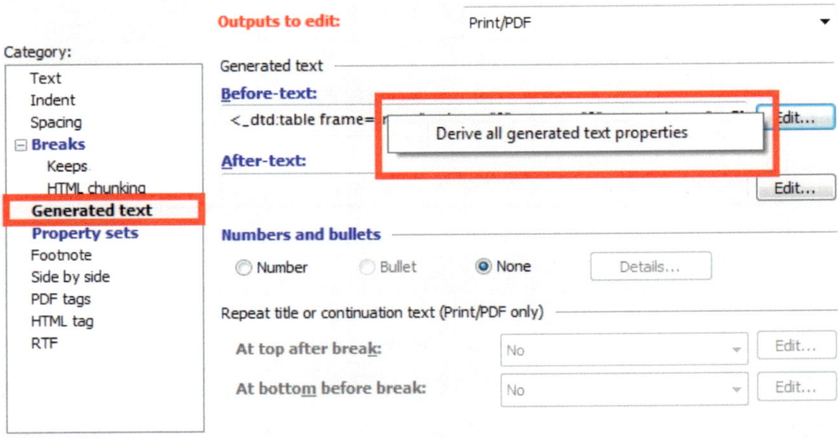

This action will remove all generated text property settings. The **Generated text** category will no longer be blue.

Select the **Property sets** category. Remove each property set in the **Used property sets** list, leaving the list empty.

If you publish PDF right now, you will find that no text is on the back of cover page anymore. That's because you removed the generated text that was generating that page. The page still breaks because the element resides in the source. Even if no content is displayed, the page break is still triggered. The next page breaks when the rendering engine encounters the **toc**, the next element that triggers a new page.

Now that the page is not being generated, you can do two things:

1. Format the **mainbooktitle** element.
2. Format the child elements of **bookmeta** so they appear on the back of the cover page.

You were halted in formatting the **mainbooktitle** because you saw the UFE and were led down the path to discover how the stylesheet was ordering processing and handling of the source elements. Now that the page is no longer being generated, you can go back and add formatting to **mainbooktitle**.

Add formatting to cover elements

Locate and select the **mainbooktitle** element rule. When you deleted the generated page content, you removed the only instance of the UFE-based context. It's a good idea to delete a context rule that is no longer in use, so that you don't spend time investigating it the next time you work on the stylesheet.

To delete the **mainbooktitle in _ufe:booktitle-meta** context rule, select it, then click **Cut** to delete it.

Select the **mainbooktitle everywhere else** context rule that remains.

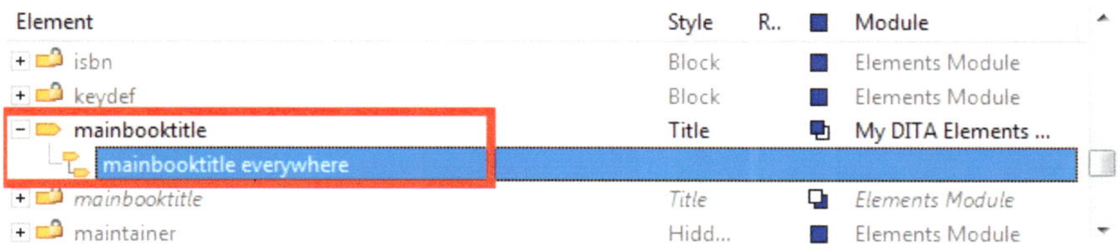

Select the **Property sets** category and add the three property sets you created earlier:

- cover title space before
- indent-left-1.5in
- indent-right-1.5in

Save your stylesheet.

Add content to the back of cover page

To restore the content on the back of the cover page, first decide what elements under the **bookmeta** element that you want to display or suppress and how you want the visible elements formatted.

This is what I'm going to have you create:

Copy the following element rules to your My DITA Elements Module:

- bookid
- isbn
- bookrights
- copyrlast
- year
- summary

Some of these elements have a style of **Hidden**. That means they are hidden by design within the stylesheet. When an element is given a **Hidden** style, there is no need to apply the hidden property set to the element or any of its children. Element rules that have the **Hidden** style applied are automatically suppressed, regardless of any individual, local, or property setting to the contrary.

If you want the **summary** to show on the back of cover page, not only do you have to unhide the **summary** element rule, but you must also unhide its parent, **bookrights**.

Element rule	**New Style**
bookid	Block
bookrights	Block
copyrlast	Block
summary	Block
year	Block

When you change an element's style type, you will see a warning:

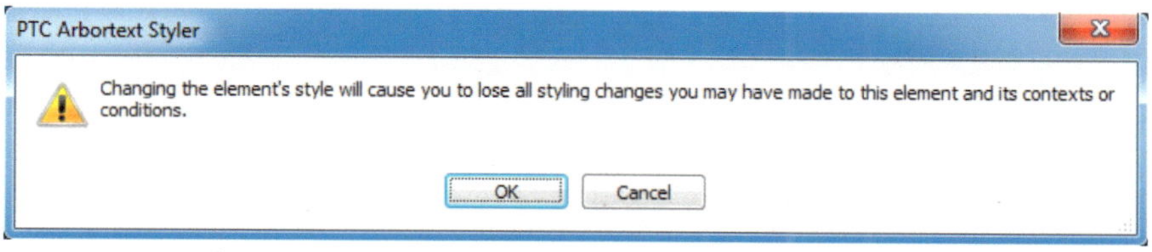

Arbortext Styler will remove all preexisting formatting settings when an element's style type changes. It will also remove any contexts and conditions. Because you want to make these elements visible, click **OK** when changing the elements with **Hidden** style to **Block** style.

The rest of the elements you copied to your **My DITA Elements Module** do not have a **Hidden** style type, but they are still not visible because they have the **hidden** property set assigned. Remove the **hidden** property set from the context rule: **isbn everywhere else** (**Output to edit**: **Print/PDF**).[4]

If you generate a PDF now, you will see that the content of the four elements show on the back of the cover page.

Format the summary

Select the **summary** element rule.

Add the following property sets:

- spacebeforemedium
- fontitalic

Save your stylesheet.

Add generated text to the copyright year

Select the **year** element rule. You can display the name of the copyright owner and the copyright year at the same time. You will use both the push and pull methods to format the copyright year content correctly.

Add the following property set:

- spacebeforemedium

Click the **Generated text** category, then click the **Edit** button next to the **Before-text** property. In the edit window, insert the copyright symbol and a space.

---

4. When you remove this property set, the output-specific format will be automatically removed. This was the only setting for that format.

Click the **Generated text** category, then click the **Edit** button next to the **After-text** property. In the edit window, type **by**, then **Insert Element Content** from the **first organization** in **bookmap**.

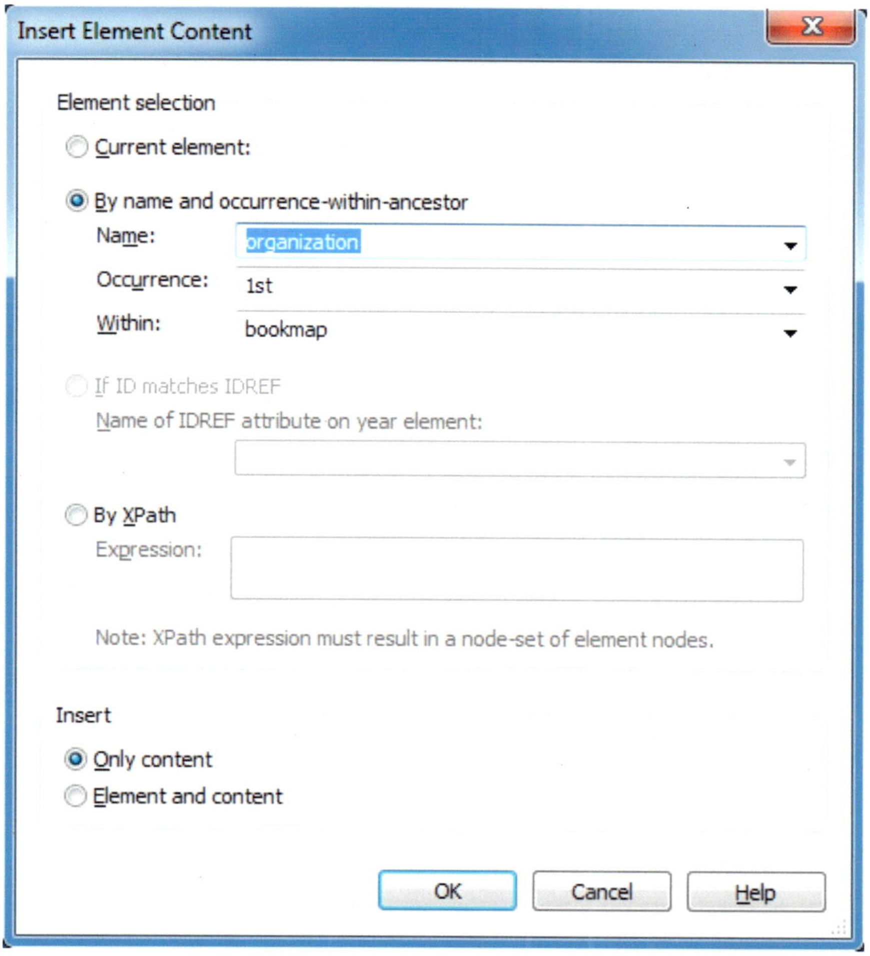

Save your stylesheet.

Add generated text to the isbn

Select the **isbn** element rule. This is the first element that appears, so adding space above it will push all visible content down to the bottom of the page.

Create the following property set in your **My Property Sets Module**:

Name	**Settings**
inside front cover space before	**Category**: Spacing • **Spacing before**: – **Preferred**: 7in – **Precedence**: high – **Minimum**: 6in – **Maximum**: 7.5in – **Keep space at top of column or page**: Keep

Add the property set:

- **inside front cover space before**

Click the **Generated text** category, then click **Edit** next to the **Before-text** property.

In the edit window, type **ISBN:**.

If you create a PDF now, you'll see the generated ISBN prefix text as well as the content pushed to the bottom of the page.

Save your stylesheet.

Chapter 10. Advanced topics

Topics Covered in this Chapter

♦ Tables of contents
♦ Stylesheet as a module
♦ Adding HTML output support

At this point, you have a basic understanding of the key components to Arbortext Styler stylesheet design. You have worked through examples of all the pillars of stylesheet design and have touched all of the most powerful tools in the designer's toolbox:

- Property sets
- UFEs
- Page sets
- Page types
- Page regions
- Generated content

These tools can be combined in different ways. Only by understanding these tools in light of how the processing works, can you fully take the next step: creating output that meets your specific requirements.

So far, the lessons in this book have provided the foundation for understanding how formatting works in Arbortext Styler. When you begin working with more complex formatting tasks, you can feel confident that you have the keys to experiment, test, validate, and ultimately address the advanced topics that end up on every stylesheet designer's desk.

Tables of contents

Tables of Contents (TOCs) are fully generated content. Because they are fully generated, you must understand how processing works, how TOCs are constructed, and how to map document hierarchy into a generated, separate representational hierarchy. A table of contents visibly demonstrates the hierarchy of all the content in the book, using indentation and formatting to represent the structure of the information.

This is what the TOC looks like in the OOTB stylesheet:

Contents

A couple of things don't look quite uniform. For example, Chapter titles have both a label and a number (**Chapter 1**), but appendixes have only a number (**A**). The indent is set at a fixed width that you might want to make bigger or smaller. You might want heading 1s (chapter titles) in the TOC to be italicized, or slightly bigger, or in a different font from heading 2.

You've already seen the **toc** element rule. You used it to trigger the **Contents Page Set**. You've also seen all the different categories of properties you can set for an element rule. While the stylesheet does have the capacity to include generated content in its **Before-text** and an **After-text** property, there didn't seem to be an obvious way to insert a fully generated table of contents. You know that you can **Insert Element** and **Insert Element Content** but you also know that you can't reliably hand-craft a TOC in an automated process.

Locate the **toc** element rule, then its context rule **toc anywhere in ResolvedMap**. We looked at this rule *and this context* once already. Now it's time to understand what they mean. The last two conditions attached to this context rule are complex and you need to examine them closely.

Working backwards, select the last condition first:

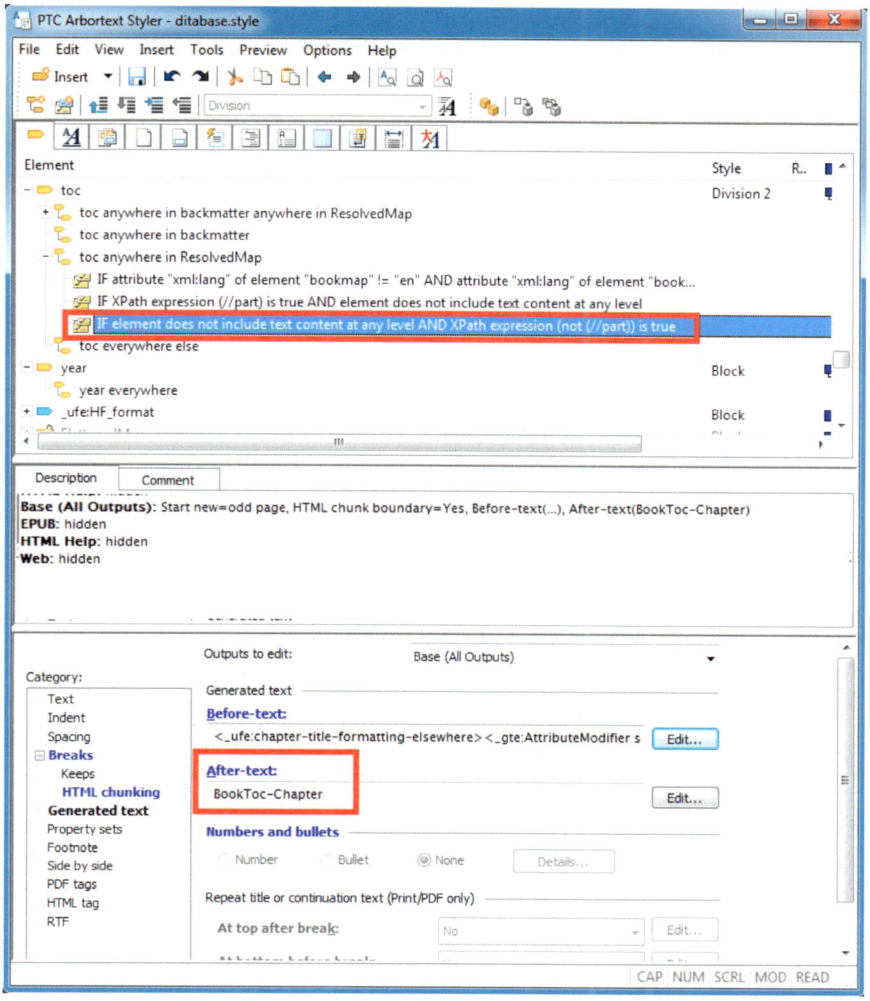

The **After-text** property of this condition, shows **BookToc-Chapter**. If you click **Edit** to get a closer look at it, you will see that it specifies a **TableOfContents**.

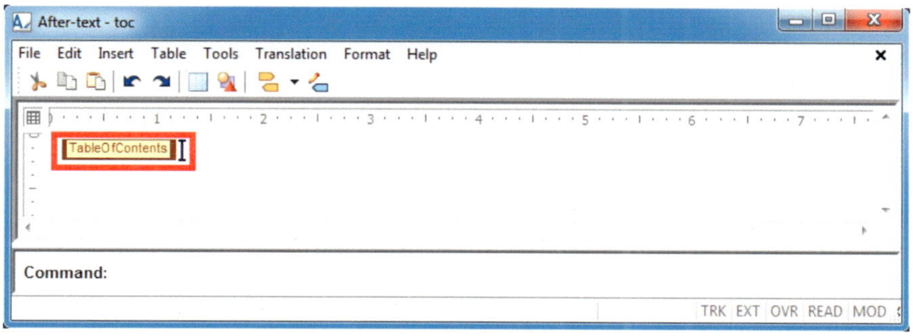

The **After-text** property showed the name of the TOC used, which is **BookToc-Chapter**.

If you look at the **Insert** menu, you see how the Table of Contents was inserted into the generated content.

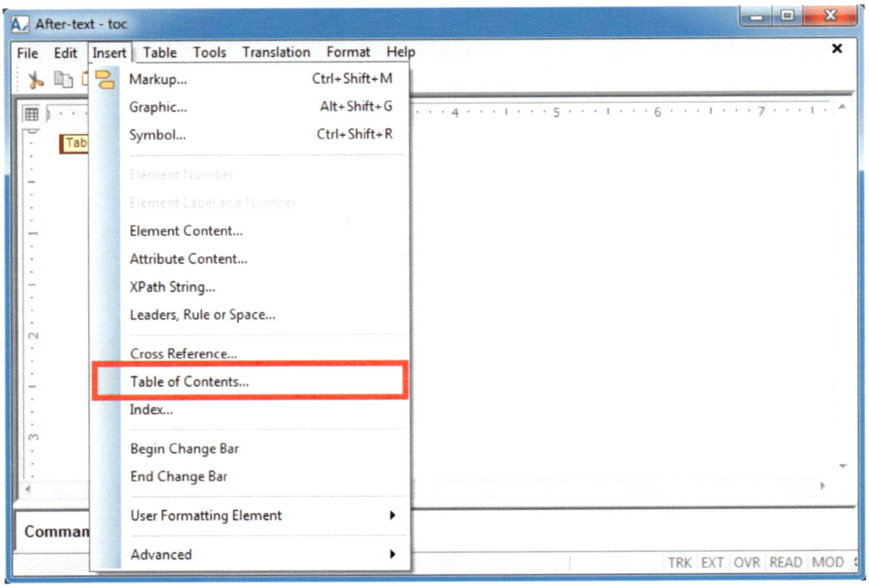

In the **Edit** window, if you look at the attributes of the **TableOfContents** element itself, you can see which TOC is being referred to in this instance.

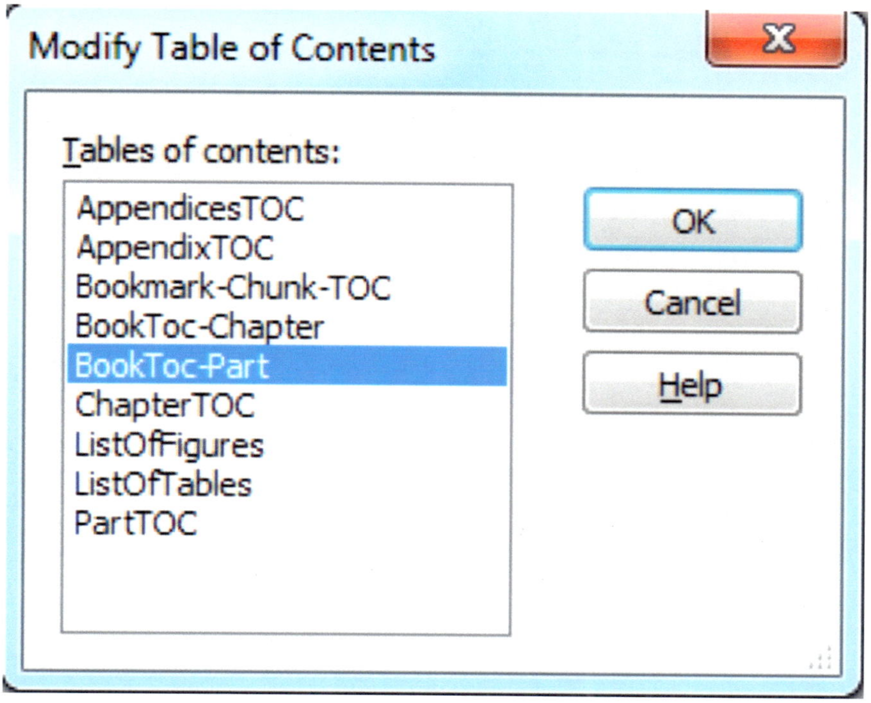

You can choose any of the Tables of Contents currently defined in the stylesheet.

If you look at the other condition (the next one higher in the list of conditions for this context rule), you will see that it shows a different TOC:

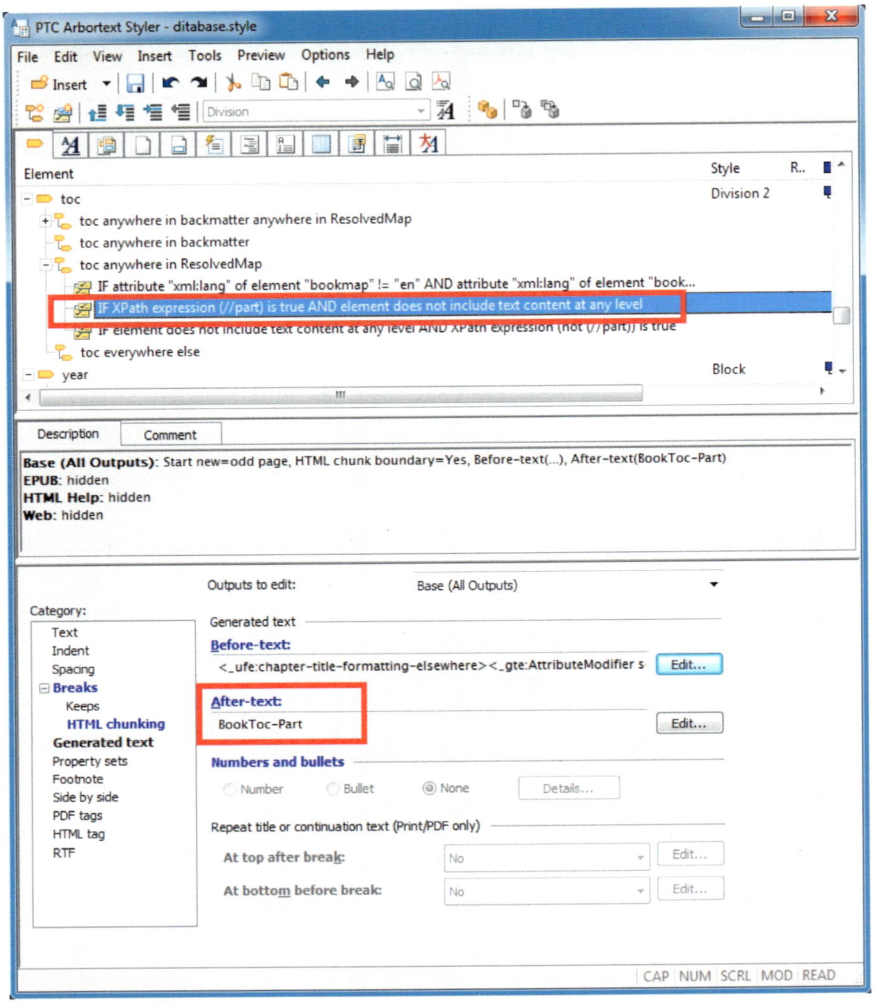

This condition invokes the **BookToc-Part**.

To find out more about both of these TOCs, go to the **Tables of Contents** tab.

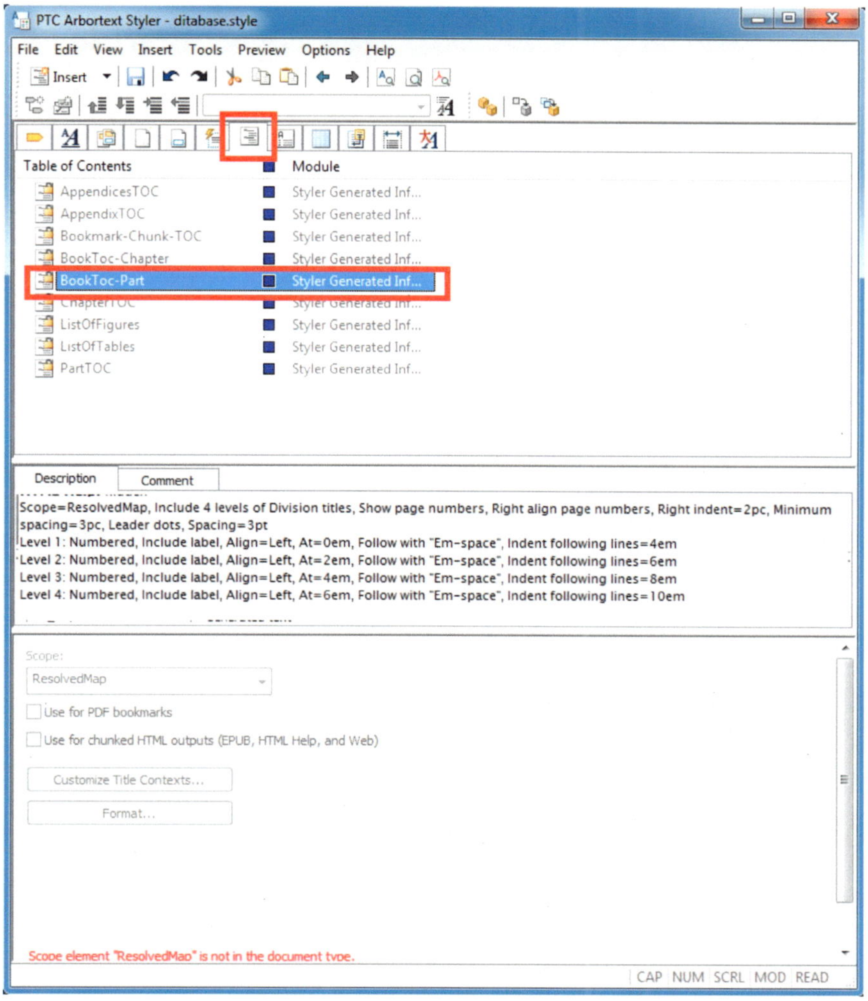

This tab shows several different TOCs. The two most important to you are:

- **BookToc-Part**
- **ChapterToc-Part**

The **BookToc-Part** corresponds to the main TOC at the front of the book. The **ChapterTOC-Part** corresponds to the mini-TOC that appears at the beginning of every chapter.

Copy the **BookToc-Part** rule to your **My StylerGenerated Module**.

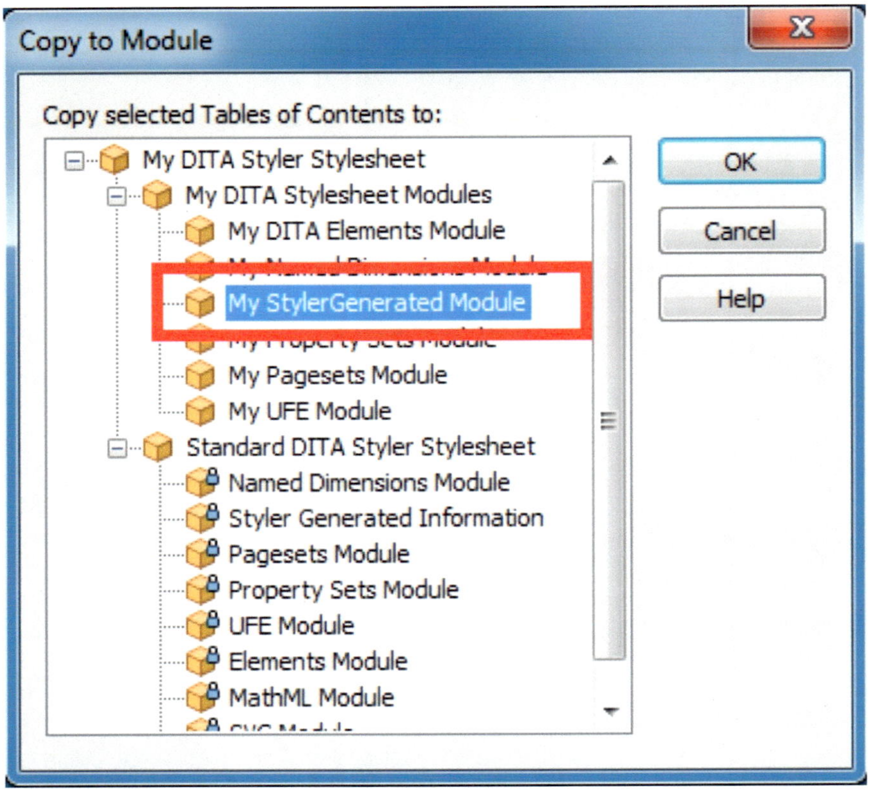

Click **OK** to return to the **BookToc-Part** rule.

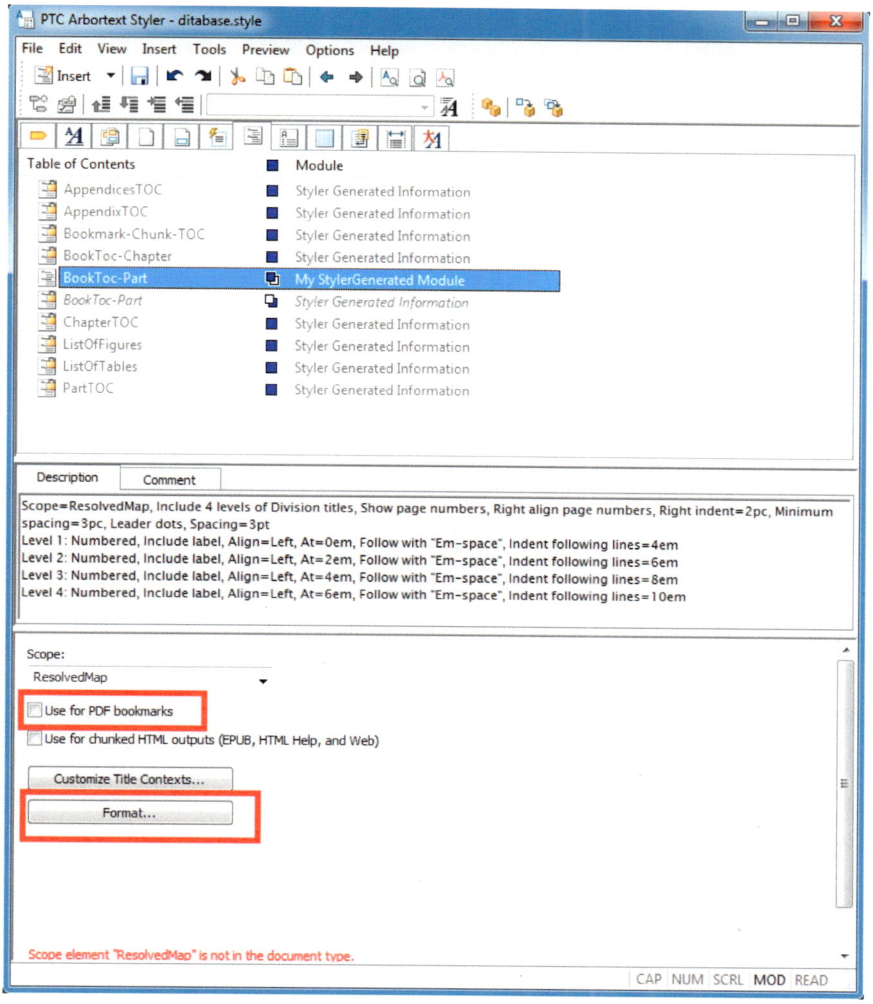

Click **Format** to open the **Table of Contents Format** window.

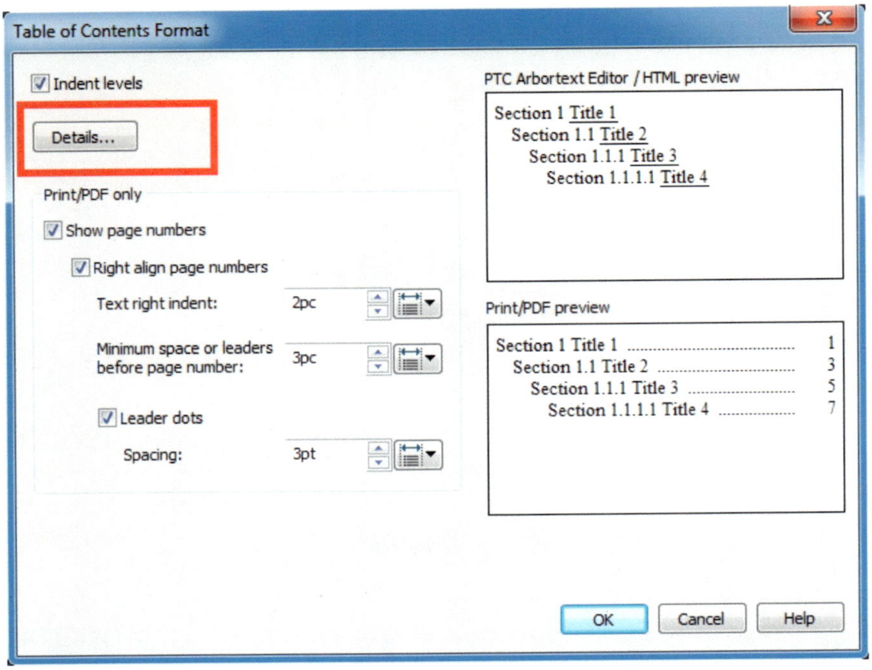

You can set basic properties here. You can establish indents and choose to display leader dots. Clicking **Details** opens the **Table of Contents Format Details** window, which gives you finer control over the formatting of text at each level.

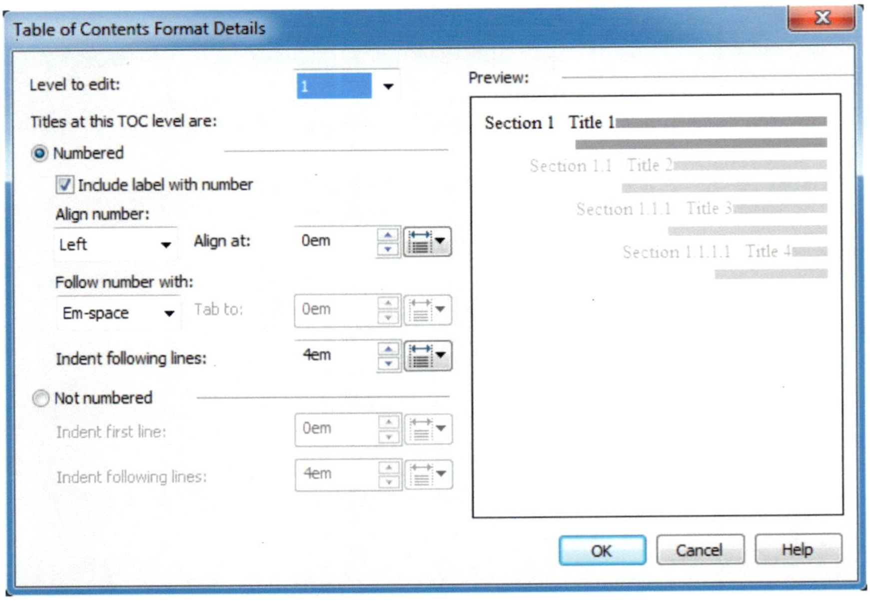

To change levels, change the value in the **Level to edit** dropdown menu.

Save your stylesheet when you're done making changes.

Stylesheet as a module

Once you have your stylesheet the way you want it, you can use it as the basis for future stylesheets.

For example, if you have a special compact document with a different cover, you can create a new stylesheet and include your current stylesheet as a module in it. To change the cover, you can create a new cover page in your new stylesheet, giving it priority by default because it exists at a higher level in the hierarchy. All you have to do is create the new cover and it will override the old cover. Because the rest of your document should look the same, once you create the new cover, you're ready to publish! Including your old stylesheet as a module in your new stylesheet means that you inherit all the rest of the formatting automatically.

Always use inheritance to your advantage!

Try an example.

Close the stylesheet editor. In the document editor, choose **New Stylesheet...** from the **Styler menu**. Give your stylesheet a title (**My Second Styler Stylesheet**). Save it as **second.style** in your **C:\Arbortext\custom\doctypes \ditabase** custom directory.

Open the modules window and add the **ditabase.style** stylesheet as a module.

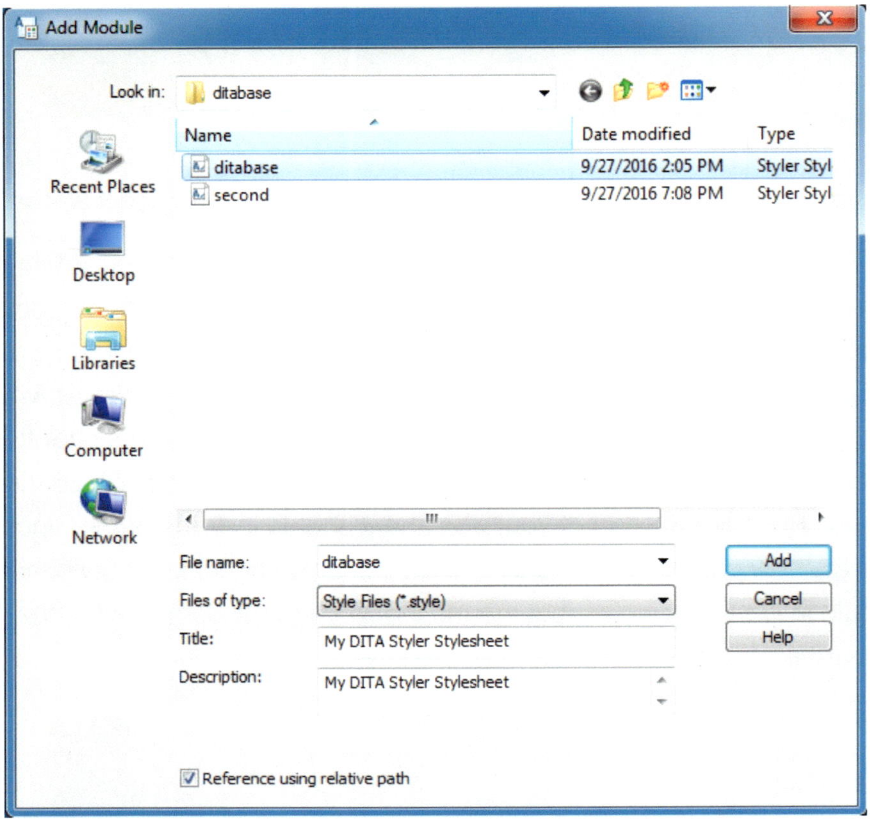

Click **Add** and see how all the modules you created are added automatically:

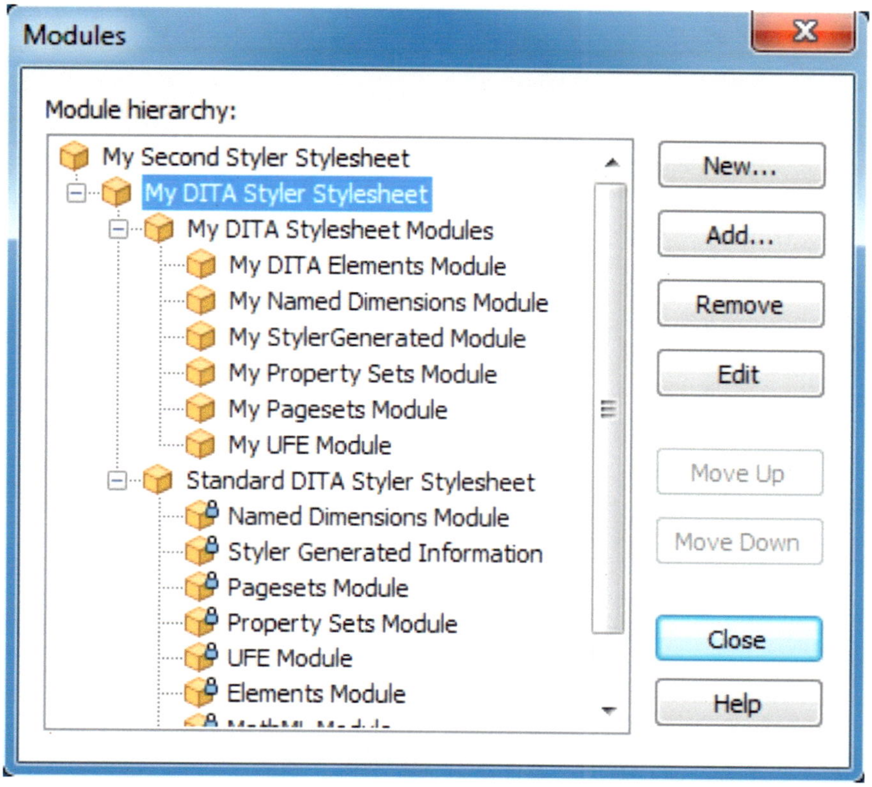

Go to the **Page Regions** tab and copy the **Cover Bottom** to **My Second Styler Stylesheet**.

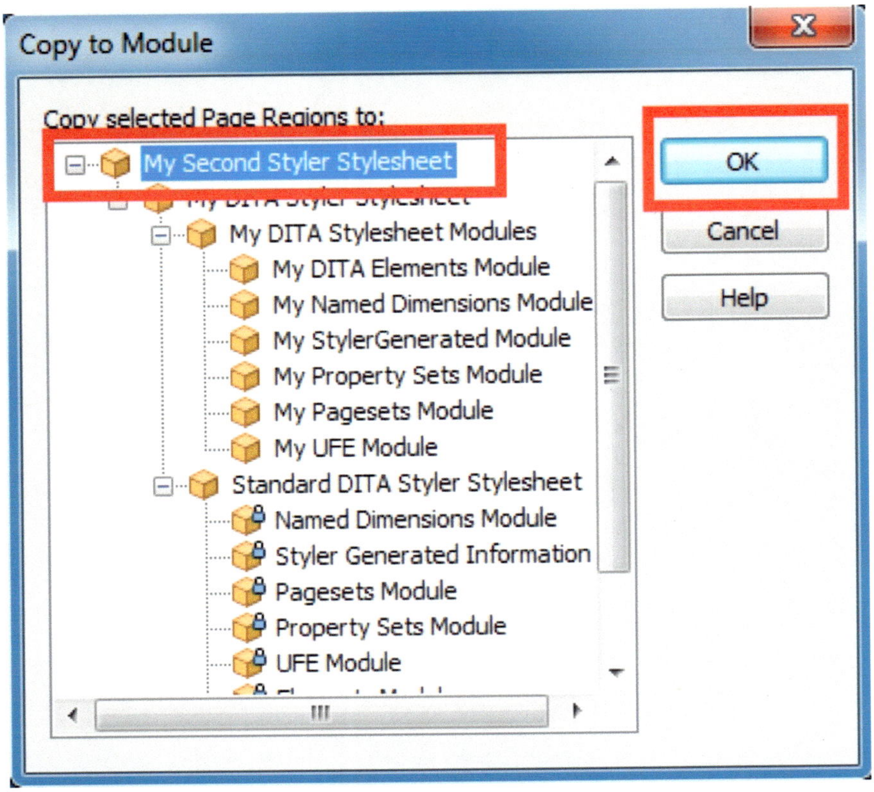

Go to the Graphic category of the Cover Bottom rule and change the graphic path to pumpkins.jpg

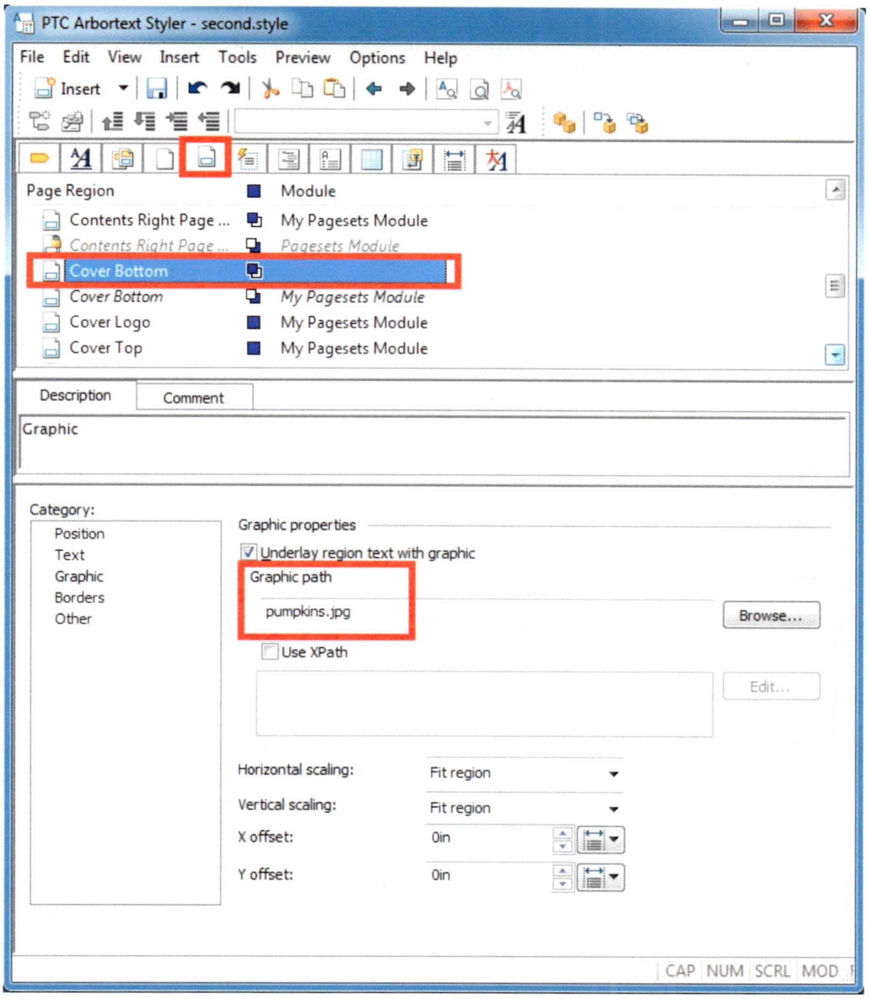

Save your stylesheet. Click the **PDF** icon in the stylesheet editor toolbar to generate a PDF.

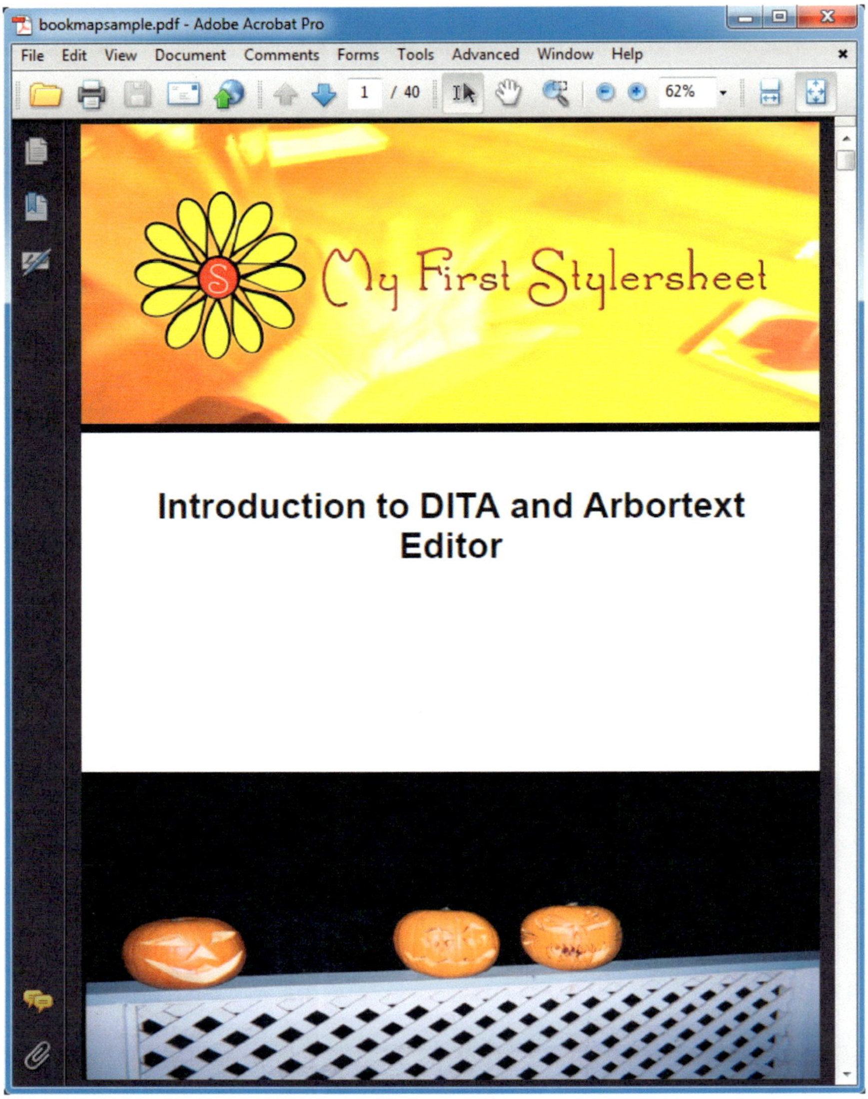

It doesn't look perfect, but we made only one small change. And we didn't change the existing stylesheet at all. In fact, if you go to the document editor and choose **File→Publish→PDF**, you have both stylesheets to choose from. You can publish with either one and get the format you expect.

This was a simple change to demonstrate how you can leverage inheritance to create new stylesheets in just a few clicks. You don't have to start from scratch unless you want to.

Expert tip!

Occasionally, you might see a warning in the **Styler Log** window that looks like this:

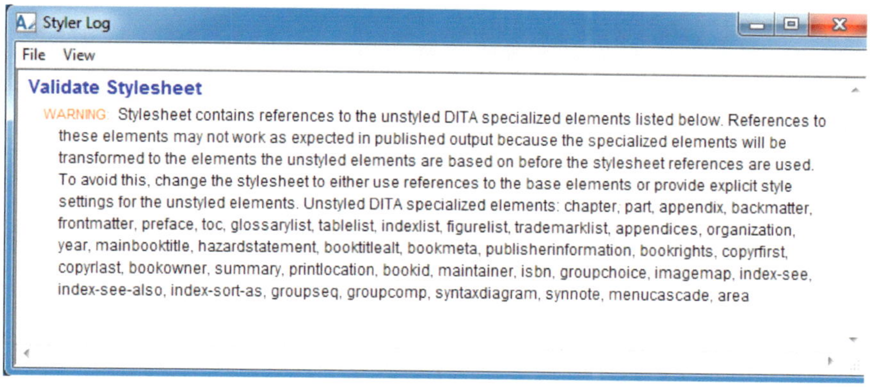

This warning is nothing to worry about. It is based on how Arbortext is reading the DITA doctype for the source document and matching the element definitions with list of element rules and their style types. It is not an error. You won't have a rule for every DITA element, especially those you don't use.

Adding HTML output support

It shouldn't be hard to figure out how to add special formatting for HTML output to your stylesheet. The first one you've already done: You used the **Outputs to edit** dropdown menu to switch between **Print/PDF** and **Base (All Outputs)**. You can do the same for HTML output.

There are four parts to formatting HTML output:

1. Output-specific formatting for one of the HTML output types
2. Properties belonging to the HTML tag category
3. Global stylesheet properties
4. The default frameset directory

HTML output-specific formatting

To change formatting for an HTML output-specific format (EPUB, HTML file, HTML Help, Web), all you have to do is change the **Outputs to edit** setting to the format to which you want to apply special formatting. Remember, you don't have to specify an output format. If you don't, your stylesheet will default to the formatting settings specified in **Base (All Outputs)**.

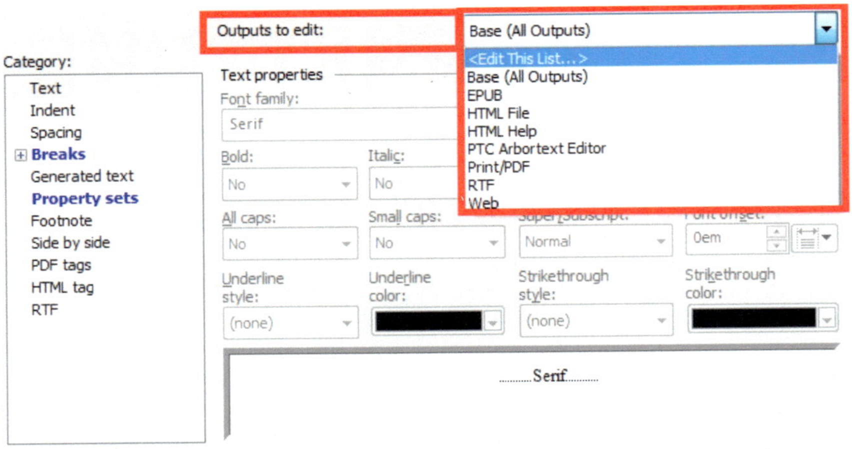

The HTML Tag category

In addition to output-specific formatting, there is a special category in the lower panel called **HTML tag**. Settings to the properties in this category apply only to the HTML output formats.

In this category, you can choose what HTML element to map the source document element to. The list of options changes based on the style type of the element you currently have selected.

Here is an example:

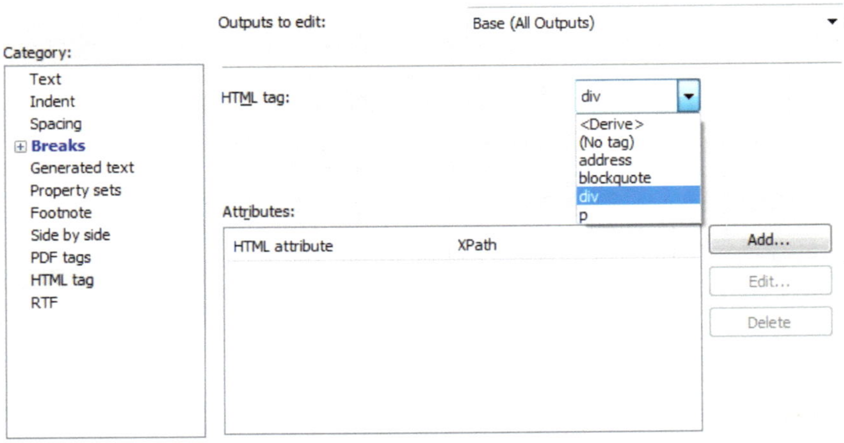

You can also specify the attribute value on the HTML element. For example, you might want to specify the type of numbering on an ordered list. Click **Add** to open the **Add HTML Attribute** window.

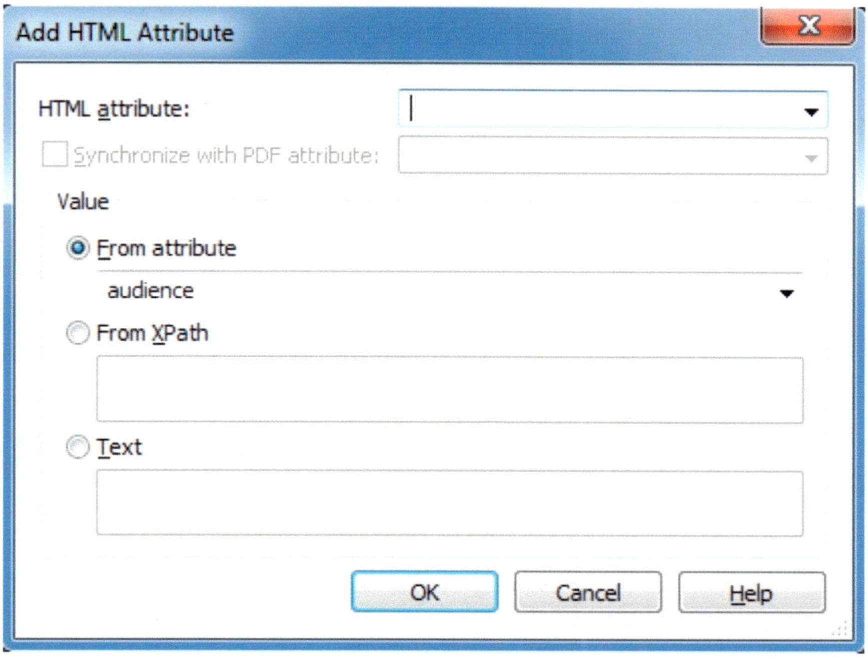

You can construct the attribute value from any element in the document.

The **HTML tag** category gives you the flexibility to customize the HTML generation and add specifics where you need them.

Global stylesheet properties for HTML output

In the stylesheet editor, choose **Stylesheet properties** from the **File** menu.

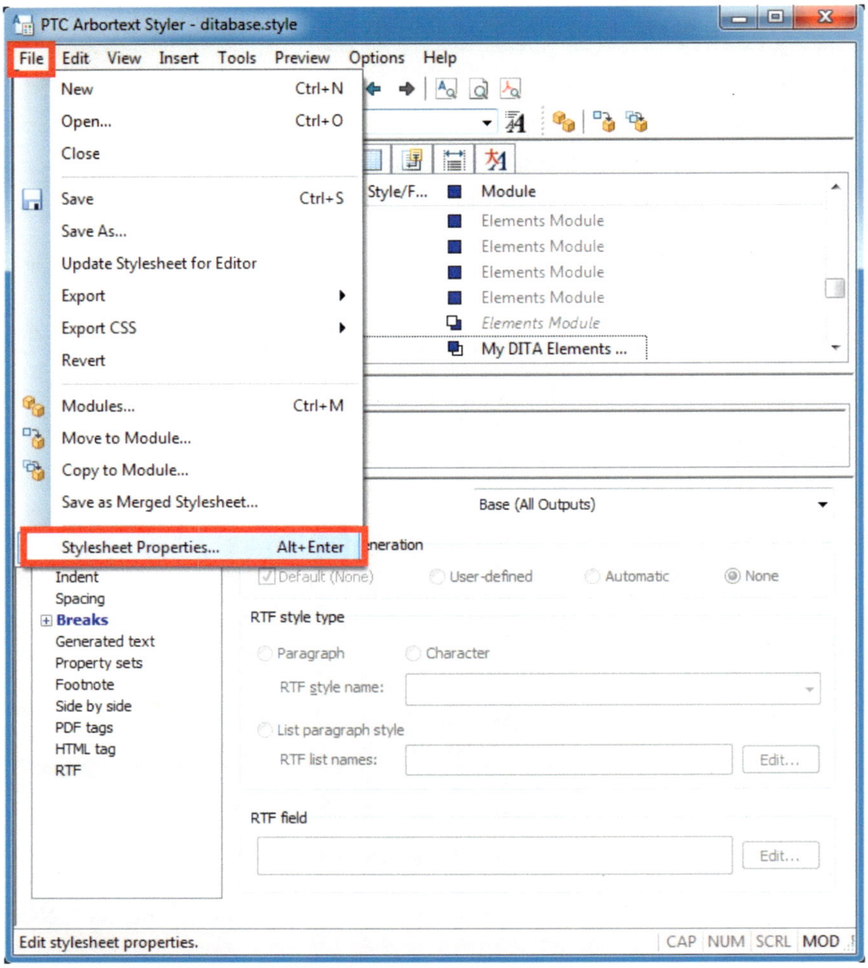

In the **Stylesheet Properties** window, go to the **HTML** tab.

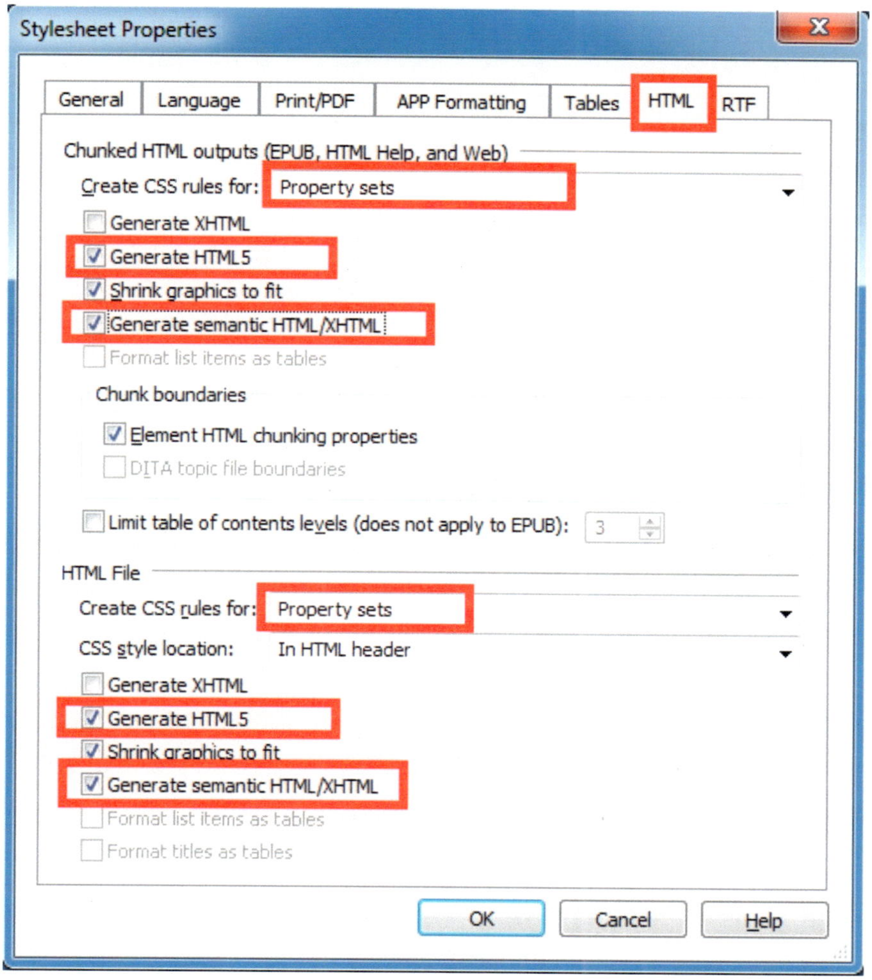

In this window, you can specify how the HTML output is constructed.

- For standard HTML5, check **Generate HTML5**.
- For lists generated as HTML list elements, check **Generate semantic HTML/XHTML**.
- To generate a .css file similar to what you are used to seeing, select **Property sets** as the option in **Create CSS rules for**.

The default frameset directory

The default frameset directory is part of your custom directory. It is the template for how HTML pages are chunked when building the **Web** output format.

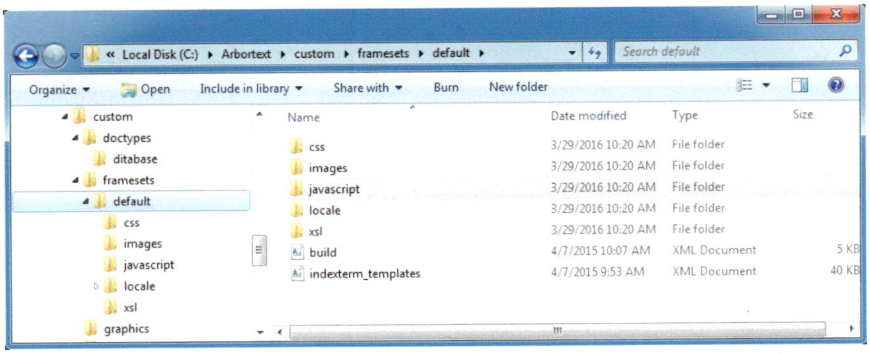

Customizing these files is perhaps the most advanced of the topics I'm going to mention. You can change any of the files in the default framesets subdirectory of your custom directory.

You can replace any of the graphics in the **images** folder to change the logo and other images to fit your corporate look and feel standards. You can also replace the files in the **css** folder. These are standard **css** files that are used when generating the **Web** output, but they are not the css files that are generated when you publish the **Web** output. Remember, you set the stylesheet properties to generate css files from the property sets in your stylesheet. The property set-based .css files do not exist until after publishing is completed.

To change some of the other files, you will need JavaScript or XSLT to make changes. These files reside in the respectively named folders here in your custom directory.

Chapter 11. Next steps

I know that I haven't covered every feature and panel in the Arbortext Styler UI —that would be impossible. These details are covered in-depth in the Arbortext Styler Help Center. However, now that you've been through this book, you should be ready to start work on your own stylesheet that reflects your formatting requirements. If you've followed the steps in this book, you've used the basic elements available to you in Arbortext Styler stylesheets. You should be able to make the changes needed to get the output format you want.

I know that using the OOTB stylesheet as a module in your custom stylesheet feels like a double-edged sword. On the one hand, it gives you a tremendous head start, providing formatting that is acceptable for a majority of technical publications. On the other hand, it serves as an example to PTC's customers of what is possible, including contexts, conditions, output-specific formatting, and unique constructions that can be challenging to track down.

Regardless, we recommend using the OOTB stylesheet as a module. While it can be take time to track down where a specific setting is configured, you learn a lot while investigating. Ultimately, the head start you gain far outweighs the challenges you face.

With the 7.0 release, the OOTB DITA stylesheets added basic formatting support for all of the MathML elements. If you incorporated the OOTB stylesheets in your 6.1 stylesheet, all you have to do is replace the 6.1 OOTB modules with the 7.0 modules, and you get all that MathML formatting for free. That's just one example. If you had incorporated the OOTB stylesheets in your 5.4 stylesheet, when you replaced them with the 6.0 versions, you got support for all the new DITA 1.2 elements.

Incorporating the OOTB stylesheets as a module is the biggest boost a stylesheet designer can get. When you add the best practice for module hierarchy that we've recommended, you mitigate the risk by incorporating them as subordinate to your own modules. The amount of work you save yourself, even considering investigative time, is too great to dismiss.

When you start a new stylesheet project, always start with the process we outlined for Chapter 3. your first stylesheet, page 19. Start fresh and create a PDF that has only the formatting from the OOTB stylesheet as a basis.

This is what the OOTB stylesheet looks like with no modification:

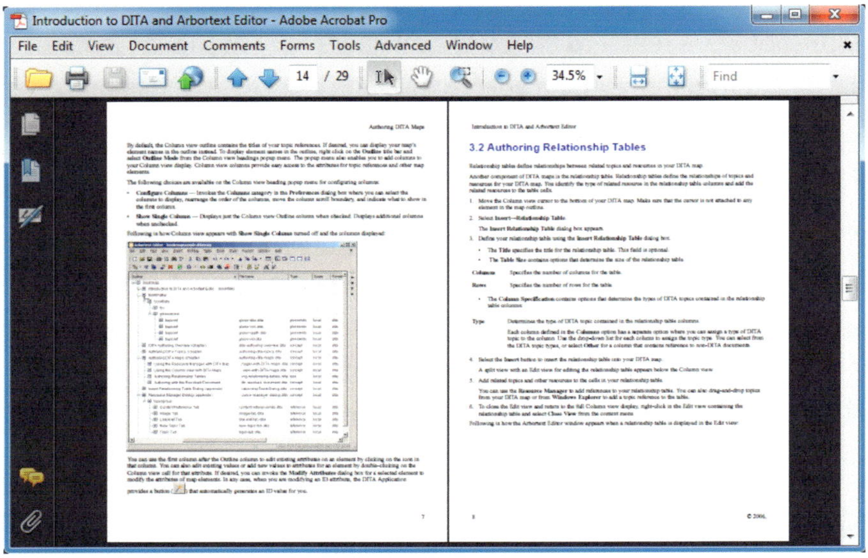

Examine each page of the default PDF to evaluate where you are and how much work you need to do.

You already know that you will likely have to create new page sets, page types, and page regions for headers and footers—and that you will have to change the generated text that appears in the header and footer. So don't pay as much attention to those details at first glance.

Pay more attention to what is showing up in the text body: what appears between the header and the footer. In the body, you find paragraphs, bulleted lists, steps, and notes, along with titles, figures (with and without captions), inline images, character formatting, and tables.

Walk through the PDF to see what works, what must change, and what you can live with (for now). For the first walk-through, note everything and rate it with one of these four categories:

Rank	Means...
Good	No change necessary
Acceptable	Not ideal, but acceptable
OK	Don't like it, but can live with it
Bad	Must change

Once you have everything identified and ranked, you have your priority list and a list of tasks to accomplish before you can declare the stylesheet good enough to start using.

It will never be perfect. When you're automating, you don't get perfect. You are no longer handcrafting each page. You're reusing content, and a page break that worked in one context doesn't work in another. You can always improve your stylesheet, but you want to be sure to prioritize so that your stylesheet isn't a roadblock to getting product out the door.

To guarantee that you have a complete list, you need a good sample document. If you have an incomplete sample document, you run the risk of being surprised at production time.

Your sample document is critical to your success. Your task list will be incomplete unless you have a good sample available when you're evaluating the default PDF output. To have a complete task list, your sample document must be representative of your content structures and the kind of information you deliver to your customers.

Your sample document should include elements you use frequently because these are the elements you need to support in your stylesheet. You can upgrade your sample, add new content, or change the content until it includes everything you want.

Be sure to save any samples you create for use later. You will want a sample that can exercise any changes you make in the future.

Chapter 12. Troubleshooting and Debugging

Topics Covered in this Chapter

♦ Add comments
♦ Use the Resolved Document for Styling
♦ Try a new sample document
♦ Clear the cache

These tricks can help you track down what's happening in your stylesheet and, in many cases, see exactly what is going on.

Add comments

You can add commentary anywhere you have generated text.

For example, if you want to test whether headers and footers are appearing as expected, you can add text that helps you recognize what's happening.

Header and footer content is easier to spot where content and page sets intersect. If you have extra characters in the different headers and footers, you will be able to see evidence of the appearance of the different headers.

When you created unique header and footer regions and generated text content for each one, you also created a situation that enables you to tell them apart. If you want to, test to make sure each one is truly separate, open them and add unique text to each one.

For example, in the **Preface Left Page Header Content**, type **(PAGE LEFT)**.

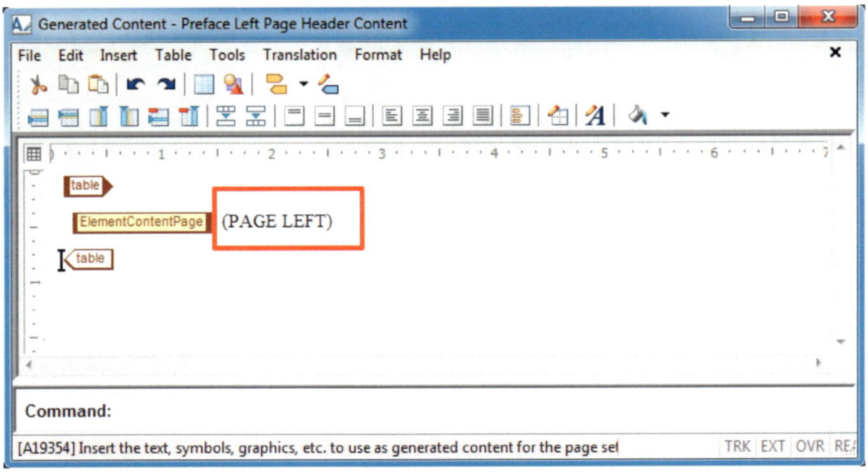

Then, in the **Preface Right Page Header Content**, type**(PAGE RIGHT)**.

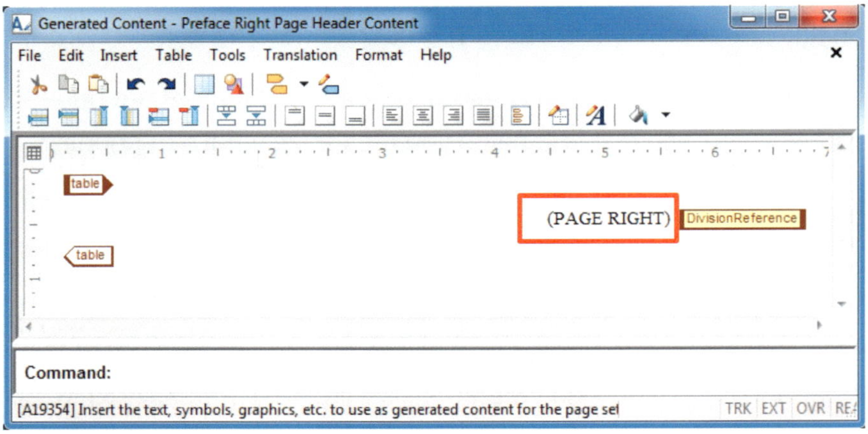

When you generate the PDF, not only will you see the Element content or the division reference, you will see the literal words **(PAGE LEFT)** or **(PAGE RIGHT)** in the header on the preface pages. You can see the page set change when the literal text disappears from the chapter pages.

It's a little trick, but a helpful one that becomes even more useful if you take it one step further.

Another trick is this one:

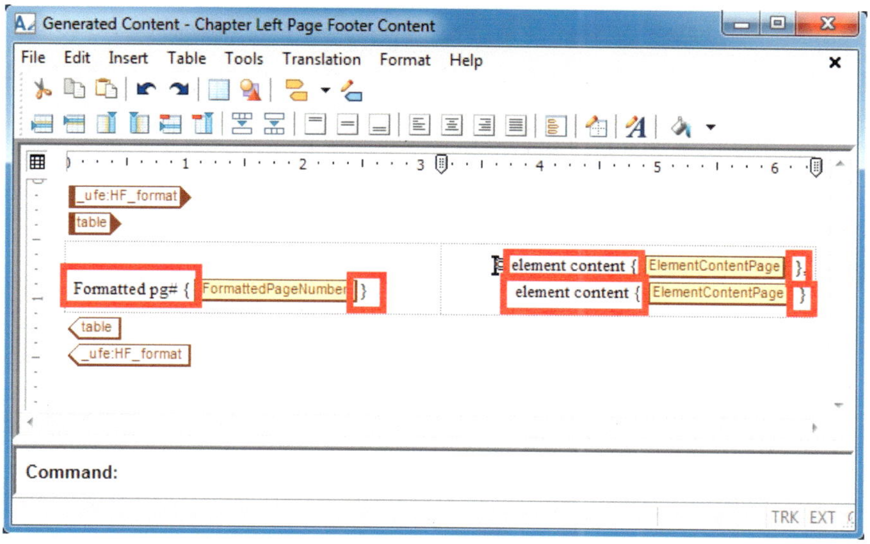

By typing **Formated pg#** or **element content** before the yellow generated text element, you will be able to see what kind of information should appear in this location (what the generated element should display). By inserting braces around the yellow generated text element, you create physically identifiable boundaries for the generated content. Without braces around the yellow generated text element, you'll never know if the element produces no value because you'll see nothing. If generated content appears between the braces, you see the characters just before and after it. If there's nothing between them, you know that the yellow generated text element isn't giving you what you expect. Augmenting generated text with literals can help you track down where the problem exists.

Use the Resolved Document for Styling

Resolved Document for Styling (RDS) is a special feature available in Arbortext Styler. The RDS will send your source document partway through the rendering engine and take all of the source content to the exact point in the publishing pipeline where the stylesheet is applied.

Being able to see what your document looks like when the stylesheet is applied can be extremely useful when you can't figure out why an element is not matching a rule the way you think it should. The element might actually be different at the time the processor is handling the elements.

Take a bookmap, for example. A bookmap has very little explicit content. Most document content resides in topics that are referenced by **topicref** elements. When you resolve your bookmap document for styling, Arbortext will open it in a separate window with all topics resolved into a single entity.

To see the resolved document for styling, go to the **Edit** menu and select **Edit Resolved Document→For Styling**.

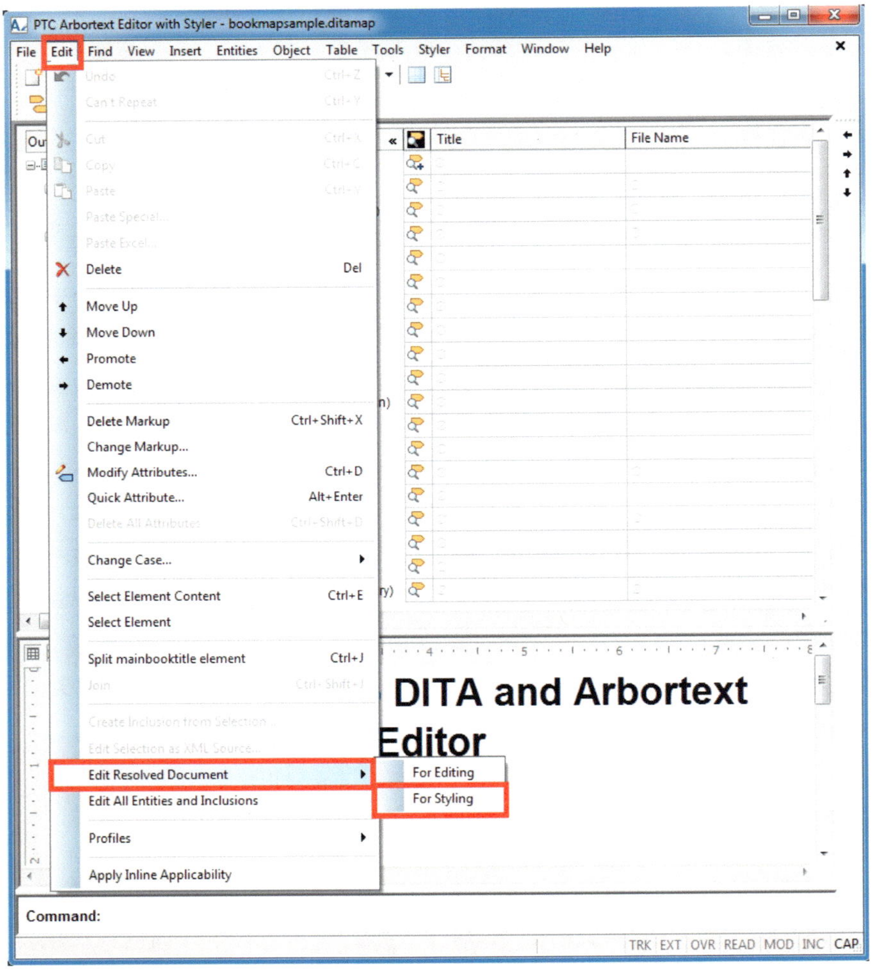

Your document will open in a new document editor window. The document will be fully resolved—all DITA conrefs resolved, all text generated (TOCs, cross references, etc.) so it is exactly the way it is before the final output is rendered.

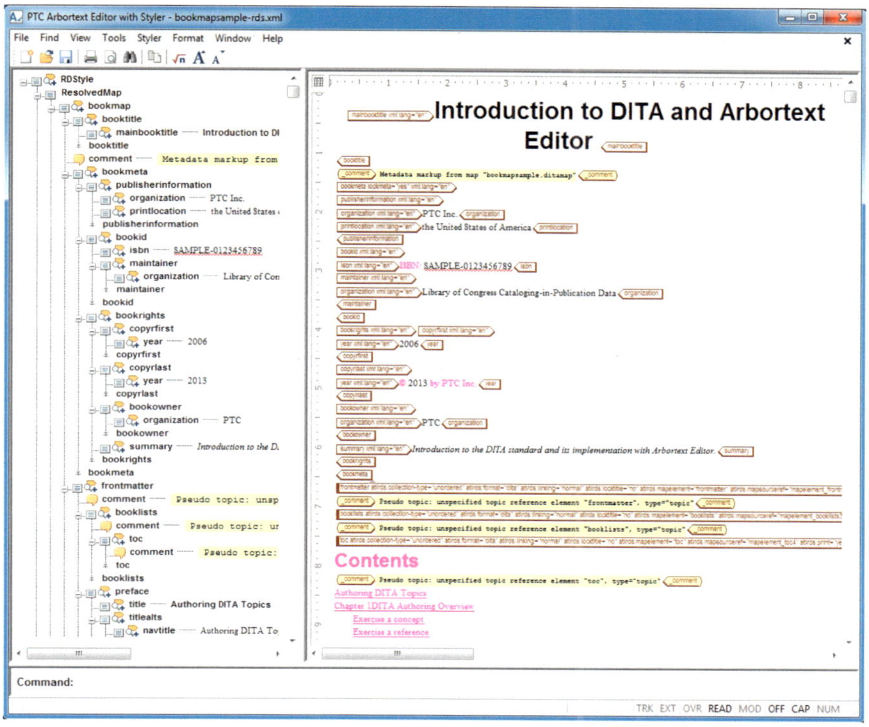

From this window, go to the **Styler** menu and choose **Open Stylesheet...**

The stylesheet editor window is always paired with a single document editor. If you want to open the stylesheet against the resolved document, you need to open the stylesheet editor from the menus in the resolved document editor window.

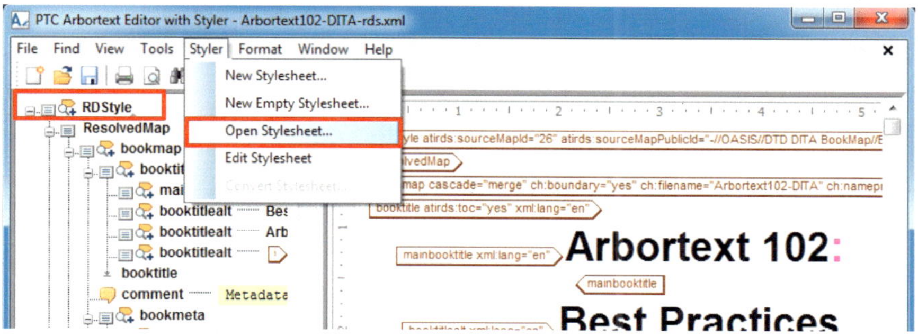

Expert tip!

A resolved document for styling isn't editable. It exists only in the document editor so you can see the full document structure as it exists. It also enables you to click in and out of the document structure to see how the elements at this point match the rules in your stylesheet.

Try a new sample document

Stylesheet design is programming. It is inevitable that you will code to your sample document. You don't code for all possibilities. The amount of work required to cover every possible situation greatly exceeds the benefit of having every possible situation covered.

Every writing team has conventions for content authoring. Either you use bookmaps or you don't. Either you use chapters or you use topicrefs. Not everyone uses topicheads. Always start with your typical case. Create a sample document that is representative.

Then expand it. Try to insert as many odd cases as possible. Look through your existing content and try to find the odd cases, the strange constructions, the difficult passages. Improve your existing sample document or create a new one that exercises the different situations in as many ways as you can expect them to occur.

Clear the cache

If you have the command line enabled, type the command **clean_cache**. Doing so will clear any cached information that Arbortext Editor or Styler has stored on your system.

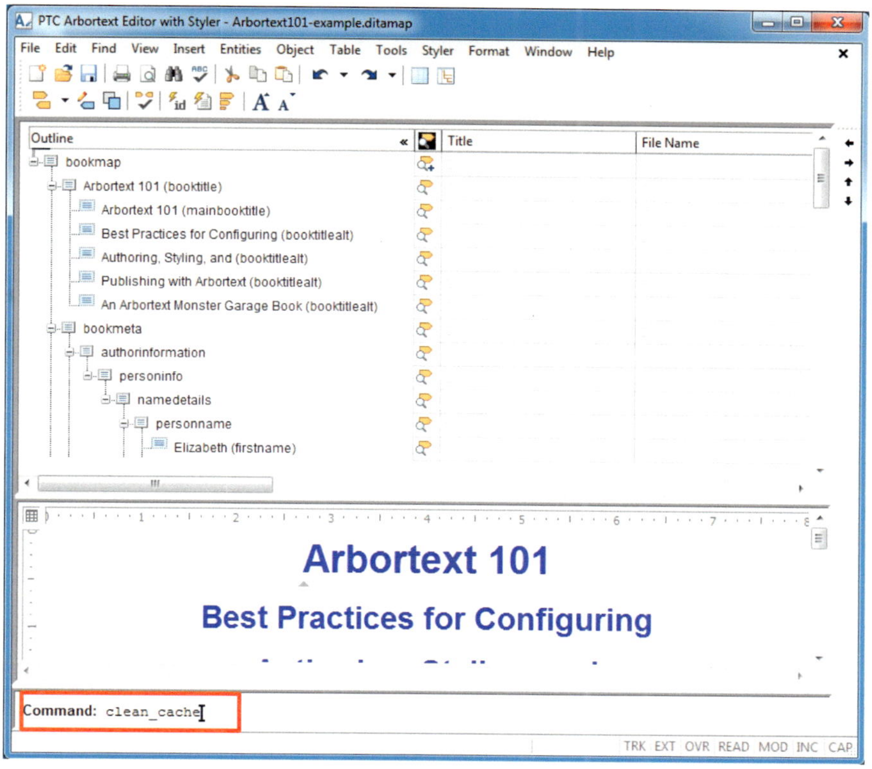

Closing and reopening the entire application can also clear temporary items stored in the cache.

Chapter 13. Where to learn more

Topics Covered in this Chapter

♦ Take the eLearning courses
♦ Programming language references
♦ Top resources for Arbortext users

You are now ready to go!

At this point, you've got enough of the basics to get started. Everything you need is in place, and the patterns I've laid out for you are reproducible.

You have a stylesheet with a solid foundation that you can now extend using the same techniques you used to set up this first stylesheet. You will find that you can create a brand new stylesheet, support a new variant for a new customer, or change branding quickly and easily, and with little effort.

Plus, you now have the basis to try advanced formatting techniques when you're ready.

Take the eLearning courses

The eLearning courses have even more value for stylesheet developers than for general authors. PTC has provided training for stylesheets from the beginning and many of the older courses have critical conceptual material for new stylesheet designers.

Stylesheet designers will want all the same courses that application users take, but they should take additional courses as well. Stylesheet developers are responsible for designing, developing, and maintaining Arbortext stylesheets. Arbortext stylesheets are used to format the screen, as well as to format paper, RTF, HTML Help, and other output formats from Arbortext Editor and Arbortext Publishing Engine.

The learning paths for stylesheet developers typically include multi-week in-person training sessions. With eLearning, you get the same training material organized and presented the same way it would be if you were taking in-person training classes with a live instructor. Instead of taking five work days away from you in a classroom, the eLearning courses break the course content into one-hour chunks that you can watch as often as you like. The flexibility of eLearning can't be beat.

Here are the recommended training courses for stylesheet developers:

- Using Arbortext Styler
- Our webinar on best practices for DITA stylesheets http://ditastyle.tcdojo.org (available on YouTube).

If you want more, use the search and change the "Version" to older release numbers:

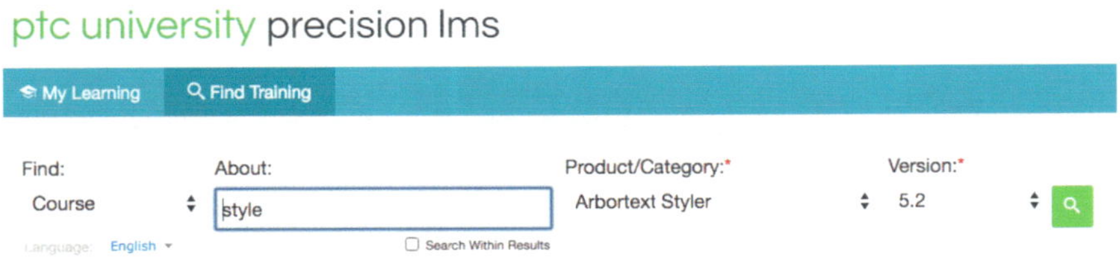

When you search, you will notice that the number of courses and their names might differ slightly. That's because every time PTC updates the software, they also update the course material. They've changed the names of the courses slightly over the years. PTC has also been building the eLearning courses one on top of the other and might assume you know the older material—or that it is old so you know it anyway.

Don't be afraid to watch an older video even if the course name is different.

Sure, the technology will be different... and, yes, the UI will be different, but you can figure it out. When you watch the older courses, you're not watching for step-by-step training, you're watching for the conceptual explanations, which is often better in the older courses.

You might even find some of the very early eLearning courses based on the Arbortext Architect product and FOSI technology. We've talked about FOSI and we've talked about Architect. It can be good to know how the old technologies and products have informed how stylesheet writers approach the task of formatting XML content. Older content will provide the context to assist you in thinking about whatever style problem you're trying to figure out.

Expert tip!

We recommend that new customers purchase a license for at least one seat of the Arbortext eLearning Library. Although the UI has changed over time, much of the earlier content is valuable because it can be more in-depth than later content.

With the release of later versions, the eLearning library course content has been updated to reflect changes to the application's UI, but much of the foundational information, because it has not changed, is not repeated in later courses. We recommend watching some of the older eLearning courses simply for the explanatory content.

Optional courses for advanced stylesheet designers

As of this writing, there is one video course about Arbortext Advanced Print Publisher (APP) in the eLearning Library: *Arbortext Advanced Print Publisher 10.0 Update Training*.

This video is not supported, as 10.0 is an old version of the APP software. It is not training in the same way that the other courses are training. This is a single hour-long video about what experienced APP users need to know to upgrade from APP 9.0 to 10.0.

Styler's base rendering engine is APP. With the release expected at the end of 2017, Styler users will be able to integrate traditional APP stylesheets and scripts into their Styler stylesheets. If you are trying a more complex layout (like spanning a table across multiple pages or synoptic alignment), you can extract the difficult tasks to an APP stylesheet rather than coding it into a source edit.

This video, while not step-by-step like the other courses, does provide some indication about what is possible and how to do it. It's also the only APP-specific course currently in the eLearning library. PTC is working to change that, and several partners have been through the first round of a new APP training course. So keep an eye out for more APP training to come because PTC is working hard to bring all the APP features to Styler developers like you!

Expert tip!

I've said this before, but for Stylesheet designers, it's never been truer: It almost doesn't matter what version of the class you take. In fact, it might be that the best version of the class in the Arbortext eLearning library is the oldest one (5.4 vs. 6.1), especially if you're watching the material to improve your understanding.

Programming language references

To learn more about XPath, we recommend the following book:

- *Definitive XSLT and XPath* by G. Ken Holman (in print or PDF).

Ken Holman has been well known as the premier trainer for XML technologies. This book is an excellent resource for learning the intricacies of the XPath. If you learn better through video, he has videos of his training courses on Udemy.

Top resources for Arbortext users

Although the Arbortext Monster Garage books are the first books about Arbortext that is not part of the software user documentation, there have always been resources available to members of the Arbortext community.

Everyone learns at a different pace. We all learn the same things. We simply do it in a different order: We learn what we need when we need it. Community resources can be invaluable if you want to tap into the hard-earned knowledge that someone else is willing to share.

Everyone learns differently. Some of us are video learners. Some of us prefer books. Some are tactile and need to touch, feel, and move things around. Some of us learn through discussion. As with Arbortext, there's always more than one way to get something done.

Here is the list of resources every Arbortext Styler user should know about:

Books about Arbortext

There are very few that are not part of the product documentation. The current list is short:

- *Arbortext 101: Best Practices for Configuring, Authoring, Styling, and Publishing with Arbortext* (http://amzn.to/2daPX9u)—the first book about Arbortext. This book explains the rationale behind why I recommended the custom directory structure I did as well as why the stylesheet fallback works the way it does.
- *Arbortext 102: Best Practices for Creating Arbortext Styler Stylesheets* (http://amzn.to/2dg5xSN)—the first book about Arbortext Styler
- —the first book about Arbortext content and Windchill.

3b2users @ Yahoo Groups

Location: http://tech.groups.yahoo.com/group/3b2users

Arbortext Advanced Print Publisher was first known as 3B2 from Advent. When Advent was acquired by Arbortext, the name was changed to reflect the product suite that it was folded into. The Yahoo group retained the old 3b2 moniker.

If you're looking for the place where serious and experienced APP users congregate, you'll want to join the 3b2users Yahoo group. This list has been around for a long time and is still active If you have advanced APP template questions, this is where to ask them. At the time of this writing, this group has more than 1500 members.

Arbortext TC Dojo @ Single-Sourcing Solutions

Location: http://arbortext.guru

The TC Dojo has a special Arbortext Edition and sponsors an Arbortext Mastermind group. The TC Dojo is a series of webinars in which the topics are chosen by the community and driven by community needs. In the TC Dojo-Arbortext Edition, we cover topics of interest to Arbortext users. Sessions are free.

The Arbortext On-Demand Channel at Vimeo provides recordings of the live Arbortext TC Dojo sessions at low cost to the public. Members of the Arbortext Mastermind group have access for free through the membership forums.

The Mastermind Group is members-only. The group meets monthly and is focused on Arbortext user issues. The group is part discussion group, part support group, part networking, and part training. It is governed by the members and has been described as, "more useful than I ever could have imagined." Currently, there are two groups in the TC Dojo that Arbortext users may find of interest: Arbortext Mastermind and Windchill Mastermind. The Windchill group is for IT staff and is more focused on their needs than those of end users. End-user Windchill issues come up in the Arbortext Mastermind group.

This group is not free. Members discuss real business issues and confidentiality is required. As a result, there is a small fee to attend to guarantee that random attendees are excluded.

Arbortext User Groups @ everywhere

The Arbortext User Group has a good presence online. There's a forum @ PTC Communities, a LinkedIn group, a Facebook Group, a Meetup Group, and a YouTube channel with copies of recordings from online meetings.

PTC Communities: http://ptccommunity.arbortextuser.org

LinkedIn: http://linkedin.arbortextuser.org

Facebook: http://facebook.arbortextuser.org

YouTube: http://youtube.arbortextuser.org

Meetup: http://meetup.arbortextuser.org

You can find Arbortext users everywhere. Sometimes you just need a little hint on where to start looking.

Expert tip!

Everything Arbortext In One Place @ HubPages

Location: http://everything.arbortextuser.org

Everything you need to know about Arbortext is condensed and explained in this one place. This page was created when the Wikipedia page on Arbortext underwent speedy deletion. In addition to historical information, this page has information Arbortext users need to navigate the PTC Arbortext channel successfully.

Arbortext Mentoring @ Single-Sourcing Solutions

Our customers want to do it themselves, but they don't want to do it alone. If you're reading this book then you, too, probably want to do it yourself.

As you develop your Arbortext technical and conceptual skills, you might still have questions, or develop new ones. No book or video can cover every situation. You can always email us (info@single-sourcing.com) if you have general questions. Getting answers to questions that are specific to you often require context and conversation. One way to facilitate that is to sign up for mentoring.

Think of going from journeyman to apprentice. You're working hard to grow your skills. You can't ask questions of a video or a book and sometimes you can't wait for days to get an answer from public forums. Our customers have found that a mentorship relationship boosts their confidence and helps them climb the learning curves faster.

Each of us at Single-Sourcing Solutions earned our expertise through carefully guided training that we received at the hands of mentors and talented coaches. Today, we are the mentors and coaches to our customers. Our mentoring services support the do-it-yourselfer as much as we support our full-service clientele.

We know that you want to take ownership of your project, your tools, and your content. Ultimately, you need to know how to maintain, grow, and extend them as your environment changes. When you work with a mentor, you don't simply outsource all that learning and responsibility. We help you to take ownership.

Arbortext Forum @ PTC Communities

Location: http://www.ptcusercommunity.com/community/arbortext

The best thing about this forum is that PTC/Arbortext software engineering staff monitors the forum. Many long-time users and tools folks also regularly monitor the forum as well. Beginning and advanced questions are encouraged. If you have a question, it's a great resource.

This forum also contains archives from the long-running Arbortext mailing list. Before it got moved to PTC Communities, the PTC/User portal hosted the long-running Adepters mailing list that had been hosted by Arbortext, Inc prior to acquisition by PTC. Originally called "adepters" after "Adept," the original name of Arbortext Editor, this mailing list has been active since 1996, five years after the product was first released in 1991. The entire archive is now contained in the Arbortext Forum at PTC Communities.

Chapter 14. Afterword

Thanks for reading. We hope the book has been useful. If you enjoyed it, please join our mailing list to get tips and resources monthly: http://mailinglist.single-sourcing.com.

This book is available through the Kindle Matching program. If you bought the print book, you can get the digital version from Amazon for only a few dollars.

While I was putting together this book my team and I kept finding formatting tricks that we wanted to include but didn't belong in this book aimed at beginners. We'll be writing these procedures up separately and offering them to our newsletter subscribers. We will also be recording a series of videos to accompany this book. Join the mailing list and be sure to hang around. Our newsletter subscribers have some real treats coming their way.

Right now, I'm working on "Arbortext 103: Best Practices for Configuring, Managing, and Publishing Arbortext Content with the PTC Server." If you want the inside scoop on using Arbortext content in Windchill, stay tuned! We'll be announcing it on the mailing list.

Everyone at Single-Sourcing Solutions participates in community projects—and we have a lot of them! To take advantage of one of our public service projects, go to http://social.single-sourcing.com and pick the one that works best for you.

You can always reach us directly, too. Email us at info@single-sourcing.com. We're here to help.

Made in the USA
Las Vegas, NV
10 November 2026

57843121R00164